PC Hardware Projects

Volume 2

By James "J.J." Barbarello

PC Hardware Projects

Volume 2

By James "J.J." Barbarello

A Division of Howard W. Sams & Company
A Bell Atlantic Company
Indianapolis, IN

PROMPT© Publications is an imprint of Howard W. Sams & Company, A Bell Atlantic Company, 2647 Waterfront Parkway, E. Dr., Indianapolis, IN 46214-2041.

International Standard Book Number: 0-7906-1109-0

Acquisitions Editor: Candace M. Hall
Editor: Natalie F. Harris
Assistant Editors: Pat Brady, Loretta Leisure
Typesetting: Natalie Harris
Indexing: Natalie Harris
Cover Design: Phil Velikan
Graphics Conversion: Terry Varvel, James Barbarello
Illustrations and Other Materials: Courtesy of the Author

PRINTED IN THE UNITED STATES OF AMERICA

9 8 7 6 5 4 3 2 1

Contents

To Kathy, Liza, and Jamie for their patience during those long nights when the Auto-XY prototype whirred back and forth, and back and forth, and ...

Preface

With your latest project designed, prototyped, and functioning properly, you have to face the task of laying out and fabricating a printed circuit board (PCB). You could have that done by a PCB fabrication house, but that's just too costly for most of us. Fortunately, with access to a CAD or "draw" program and a laser printer, you can lay out and fabricate a professional quality PCB. Wouldn't it be nice if your computer could also help you with the laborious and error-prone task of drilling the component holes in that PCB?

Stepper motors are readily available today at relatively low cost, and most of us have a "spare" personal computer (or PC) lying about the house. Put those two together with some very smart software, and you can easily "computerize" the PCB drilling process.

This book begins with an overview of stepper motors. It discusses what they are, how they differ from conventional and servo motors, and how to control them. It investigates different methods to control stepper motors, and provides you with circuitry for a dedicated IC controller and a discrete component hardware controller.

With this foundation, you begin construction of an automated, PC-controlled PCB drilling machine that can drill as many as 500 holes in a 6 x 8" PCB with excellent repeatability and resolution. Called *Auto-XY*, it uses readily available electrical components, hardware, and construction materials — no special purpose ICs or precision machined parts are used to create this precision device. In addition, *Auto-XY* doesn't require any extraordinary machinist or carpentry skills, but is ripe for customization based on your individual capability. Best of all, *Auto-XY* can be constructed for under $100 (hundreds less than the lowest cost commercial unit). A full-featured software application (on the companion floppy disk) rounds out the project.

With construction completed, you'll learn how to create the data file needed to drill a PCB, or how to use the Excellon drill file from your favorite CAD program to automatically create the *Auto-XY* drill data file. You'll then walk through an actual design layout, creating a PC design and board using two methods (manual using a Resist pen, and "automatic" using a commercially available Toner Transfer System). Finally, you'll see how the drill data file is determined from the layout, and then drill the PCB using *Auto-XY*. The book rounds out with a discussion of some possibilities for further enhancement of the project.

So, if you've always wanted to work with stepper motors and need a rig to drill your homebrew PC boards quickly and accurately, get out your soldering iron and follow along in *PC Hardware Projects, Volume 2*!

A Collection of Motors

CHAPTER 1
Making Things Move

Introduction

As an electronics hobbyist, you're familiar with the world of integrated circuits (ICs) and such. However, there's another related world out there. It's the world of electromechanical gizmos that buzz, sense, and make things move. If you're like me, that world has always fascinated and, at the same time, intimidated you. Oh sure, you've powered up a DC motor from a set of batteries. You may have even tried your hand at using motors to create your own robotic thingamabobs. However, as you've experimented, you found that motors in general aren't very precise when it comes to controlling the position of things. Factors like inertia, load, drag, and speed make precisely positioning something with a regular motor all but impossible!

Then, one day, you're sitting at your workbench and your mind wanders to that time when you were having lunch at the hamburger joint. At the next table, you overheard two machinists talking. They were saying something about "servos" and "steppers," and delighting in the fact that this one or that had so many degrees per step. As you came back to reality, you remembered some material you read years ago. You eagerly waded through your library of electronics articles (we all have one, right?) and found the article you remembered. "Oh, I get it," you say to yourself, "Servos are motors... sort of a precision, small AC motor; and steppers are stepper motors... sort of like a servo, but powered by a DC." While probably fascinating, the article didn't quite provide enough "how-to" for you to actually make something. So, like the first time you read it, you put the article back into your electronics library and thought, "Some day, I'll do a project with steppers."

Motors

Well, that day is here. Let's begin our journey into actually making something move, with a practical look at stepper motors. To start out, a stepper motor is simply a specialized DC motor. Both a stepper and standard DC motor have a rotor (the rotating thing connected to the shaft), and a field winding. The rotor contains permanent magnets. The field winding has one or more coils (electromagnets) that are energized with a DC voltage. Those electromagnets create a magnetic field that opposes the magnetic field of the permanent magnets. This is what causes the rotor to rotate, or at least to start to rotate.

If that was all there was, the rotor would rotate until the north poles on the rotor aligned with the south poles of the field coil. Then it would stop, firmly held in place by the magnetic attraction. Clearly, we need something else. In a DC motor, that "something else" is a commutator. The commutator is a cylinder electrically isolated into several axial segments and mounted on the rotor shaft. Voltage is fed to the field coils, from the outside of the motor, through two carbon brushes that contact the commutator. As the rotor/commutator combination spins, alternate portions of the commutator are fed by the external voltage. This has the effect of continuously varying the polarity of the DC voltage, and thus, the polarity of the magnetic field in the field coils. So, as the north pole of the rotor magnets approaches the south pole of the field coil, the voltage is reversed and the field coil switches to a south pole. This repels the rotor and keeps it spinning.

The problem with this is that you don't know which field coils the commutator is contacting at any given time. Even if you did, your hands aren't fast enough to turn the voltage on and off, to start and stop it wherever you want. Even further, if you could, the inertia of the rotor would keep it moving (unless you were able to precisely switch the voltage in the right field coil, to cause a magnetic attraction that would serve as a magnetic break).

The Stepper Motor

The stepper motor solves all these problems. Instead of a commutator, the stepper motor has field coils that are fed by an external switching circuit. The external circuit could be a set of transistors, a specialized IC, or a series of relays.

There's another difference between a DC motor and a stepper. As previously indicated, the rotor in a DC motor is placed into motion by opposition to magnetic fields. That is, as the rotor gets to a certain point, the field coil's magnetic pole is made the same as the rotor, *pushing* it away. In a stepper, the opposite is true. The voltage is switched so the rotor is *pulled* toward the field coil. Once it gets there, it's held in place as long as the voltage is applied to the field coil. To get the stepper's rotor to the next position, you need to energize the appropriate field coils to "coax" the rotor to the next position. Another nice feature of a stepper is that you can "pull" the rotor in either direction. So, with the proper control signals, you can step the rotor in a clockwise or counterclockwise direction.

The precision of a stepper motor is controlled by its physical design. The more permanent magnets it has, the smaller the distance that the "pull" will be to the next. In a stepper motor, the pull from one position to the next is called a *step*. A typical stepper motor will have 7.5 degrees per step. This means that each time you "pull" the rotor to the next position, the shaft will have turned 7.5 degrees of its possible 360 degrees (a full circle). Therefore, this typical stepper motor will have to make 48 steps to complete one full revolution. Since the position of each step is controlled by the physical design, the rotor will stop in the same position each time. Now that's precision!

Bifilar vs. Bipolar

There are a lot of different types of stepper motors (just like there are a lot of different types of ICs). However, they basically fall into two major categories: bipolar (*Figure 1-1*) and bifilar (*Figure 1-2*). A bipolar stepper motor has two (bi) field windings. A

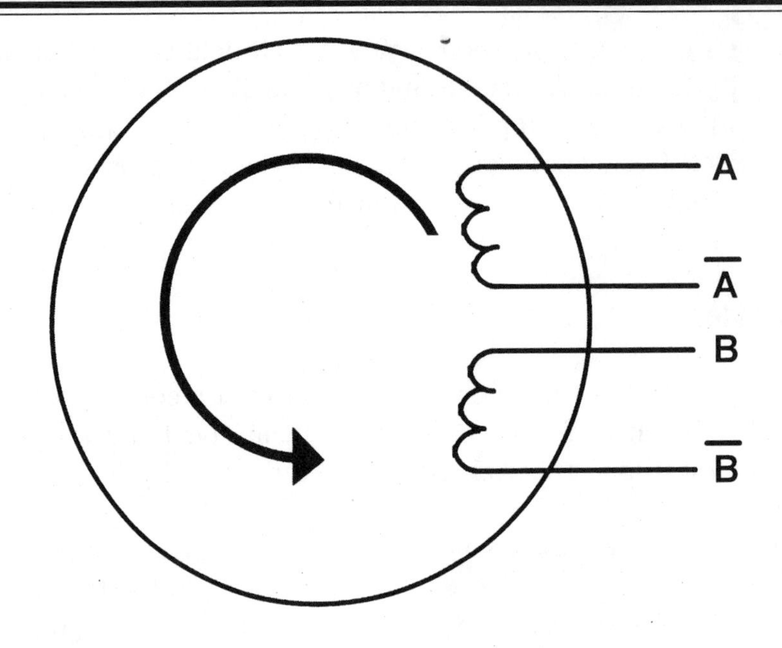

Figure 1-1. *A bipolar stepper motor.*

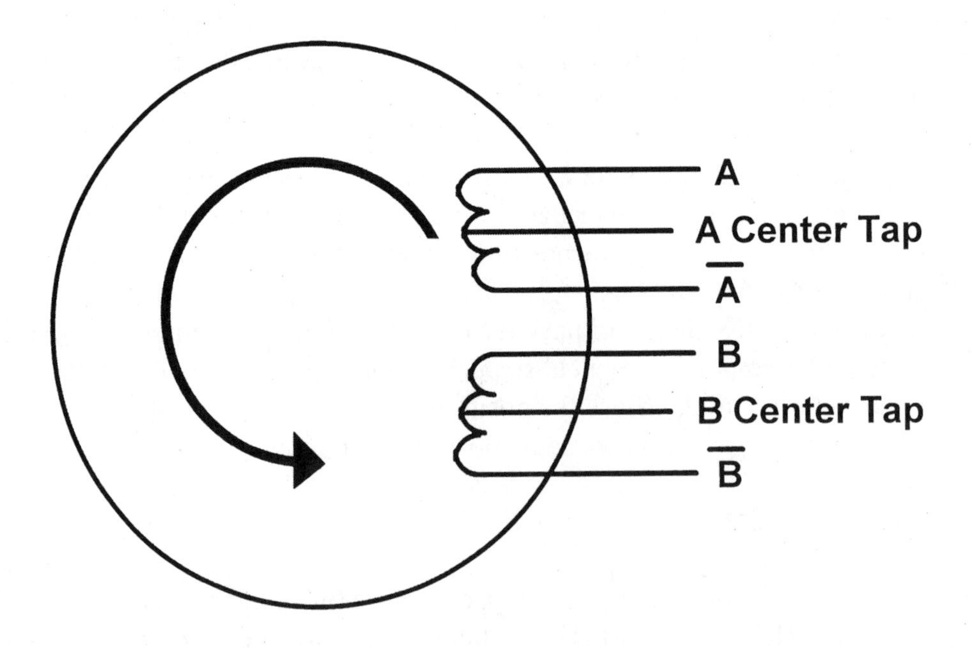

Figure 1-2. *A bifilar stepper motor.*

bifilar motor also has two field windings, but each of the windings is center tapped. So, a bipolar stepper motor usually has four leads (one for each end of each of the two windings) while a bifilar stepper motor has six leads (one for each end, plus one center tap for each of the two windings). While this does not seem like much of a distinction, it makes a big difference in how to control the voltage to the field windings. Chapter 2 gets into more detail on how to control each type. For now, I will end this discussion by saying that, for various reasons, the projects in this book will use bipolar motors. However, throughout the book, I will always describe how a bifilar type stepper could be used in lieu of a bipolar stepper motor.

The Project

Take a stepper motor and attach its shaft to a threaded rod. Then attach a threaded nut to the rod and secure it from rotating. As the stepper motor turns, the nut will travel linearly along the rod. Let's call that group of stepper, rod, and nut a *linear controller*. Now, take another linear controller and position it at right angles to the first. You have the basics of an X-Y machine.

If you attach a table to one of the linear controllers, and a mini-drill to the other, you can precisely locate any X-Y position on a flat surface. Place a printed circuit board (PCB) blank on the table, and install a small drill bit in the mini-drill, and you can precisely locate and drill holes in the PCB. That's what our project, *Auto-XY*, is all about. The great thing about it is that you don't have to be a master mechanic to build one!

Let's Begin

By now, you're probably itching to get some grime under your fingernails. So, start searching your parts sources for a few 2-phase bipolar stepper motors, with a 5 volt coil drawing less than one amp. With that said, let's begin!

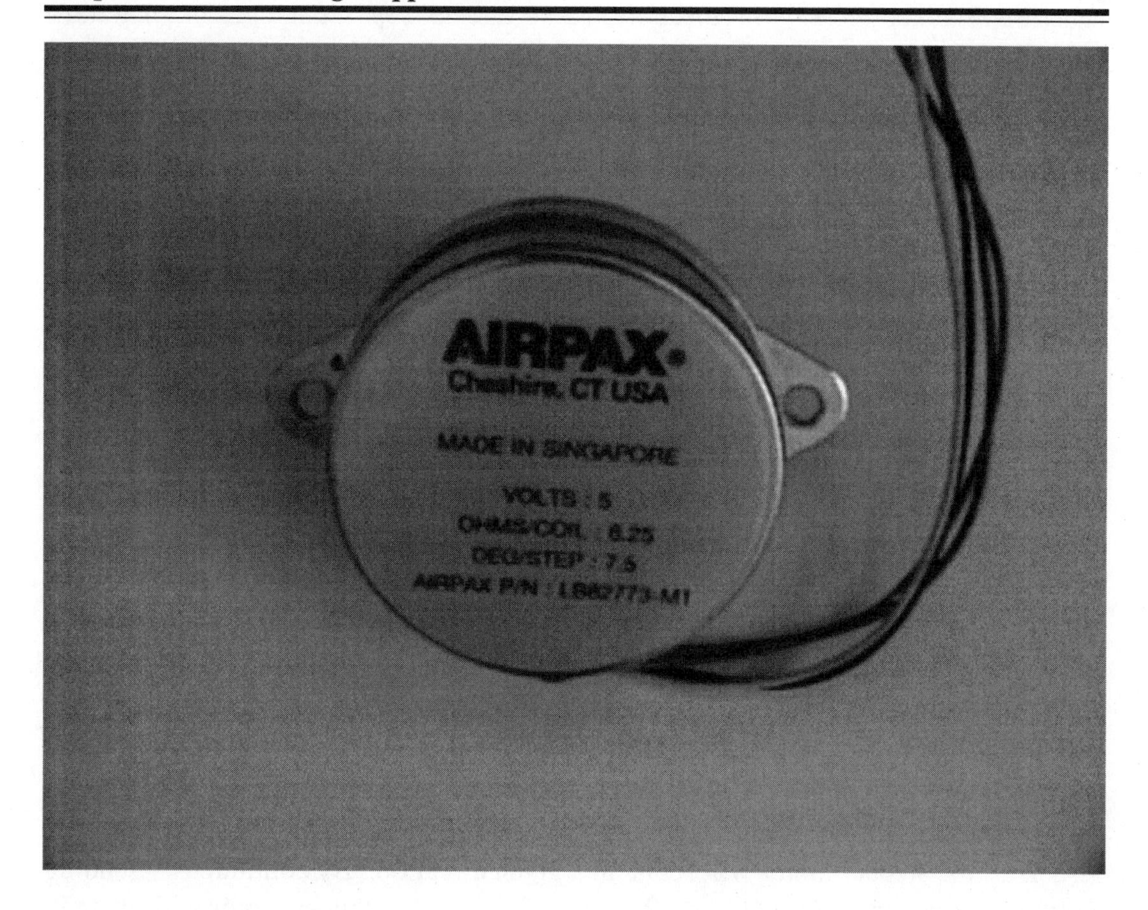

CHAPTER 2
Controlling Stepper Motors

Introduction

In Chapter 1, we introduced stepper motor basics. In this chapter, we'll get very specific on how to control a stepper motor. We'll show how to do this with simple single pole double throw (SPDT) switches. We'll also present a transistor-based controller that is fed by a computer parallel port. A third approach, using an application specific IC, will be discussed (although the IC alone cannot supply the magnitude of current required by the stepper motors we'll be using). As we investigate each of these approaches, we'll describe the modifications needed to control a bifilar (rather than bipolar) stepper motor.

The Control Sequence

What's a Sequence?

When you apply a DC voltage to a regular motor, the magnitude of the voltage determines both the motor's speed and the torque. A stepper motor's torque is also affected by the voltage magnitude, but the speed is independent of voltage. In fact, a stepper motor is sort of like a digital device. The voltage to the coil leads is ether full ON or full OFF (sort of like the one - zero status of digital bits). The magnitude of voltage applied to each end of each coil can be represented by a "0" or "1." The representation of that status is called a sequence step. Put a number of those steps together and you have a control sequence. The specific zeros and ones in the sequence determine if the stepper motor rotates clockwise, counterclockwise, or just sits there chattering.

A Standard Sequence

Most 2 phase bipolar stepper motors will respond to a standard control sequence. (*Figure 2-1*) Note that the sequence for step 1 has both coils energized (one end of either coil is energized and the other end is grounded). Each succeeding step reverses the polarity of applied voltage on one of the two coils. Also note that after four steps, the sequence repeats. By continuously sending out the four step sequence (1, 2, 3, 4, 1, 2, 3, 4, 1 ...), the stepper motor will rotate in a given direction. By continuously sending out the sequence in reverse (1, 4, 3, 2, 1, 4, 3, 2, 1 ...), the stepper motor will rotate in

STEP	A	A*	B	B*
1	1	0	1	0
2	0	1	1	0
3	0	1	0	1
4	1	0	0	1
1	1	0	1	0

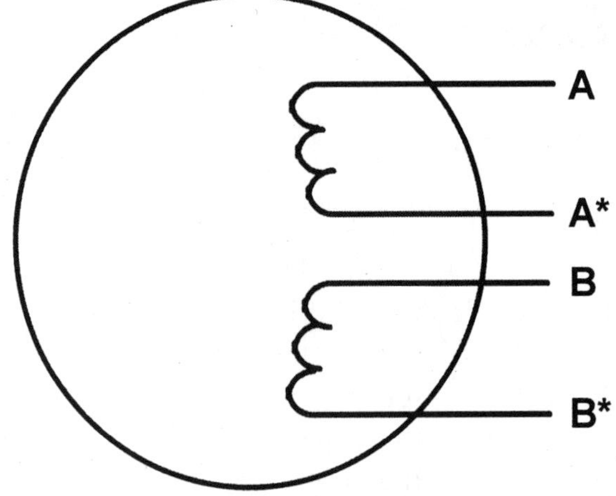

Figure 2-1*. A standard control sequence for a 2 phase bipolar stepper motor.*

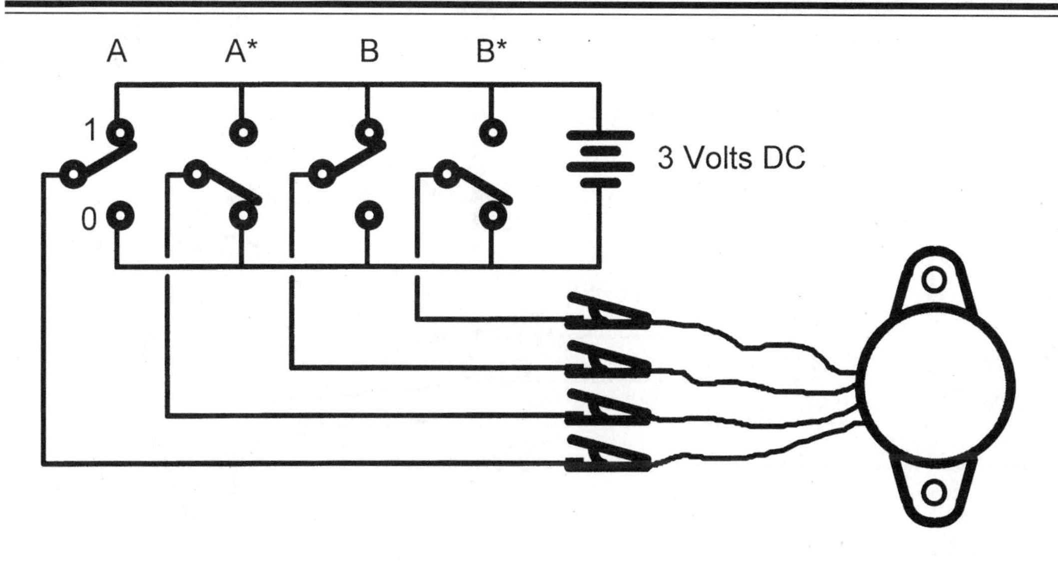

Figure 2-2. *A simple test setup containing four SPDT switches.*

the opposite direction. The time between the sequence steps determines the speed of rotation. That time can be short, producing a seemingly continuous rotation, or long, producing a series of discrete steps. If the time is too short, the stepper motor will not be able to physically respond, and it will not rotate. Instead, it will produce a buzzing sound as the motor attempts to respond, but can't.

A Test Setup

The simplest test setup is shown in *Figure 2-2*. It contains four SPDT switches. As shown, the switches are wired to the positive and negative terminals of a 3 volt DC battery pack (two 1.5 volt DC "D" cells in series). The position of the switches matches the step 1 sequence of *Figure 2-1* (1 - 0 - 1 - 0). The center lug of each switch is wired to a lead terminated with an alligator clip. Each alligator clip is, in turn, connected to one lead of the stepper motor under test.

In order to determine which switch leads should be connected to which stepper motor leads, follow this simple procedure:

1. Using an ohmmeter, identify two leads that show a very low resistance. Arbitrarily call these the two leads for coil A. Arbitrarily call the other two the leads for coil B (verify a low resistance between these leads as well).

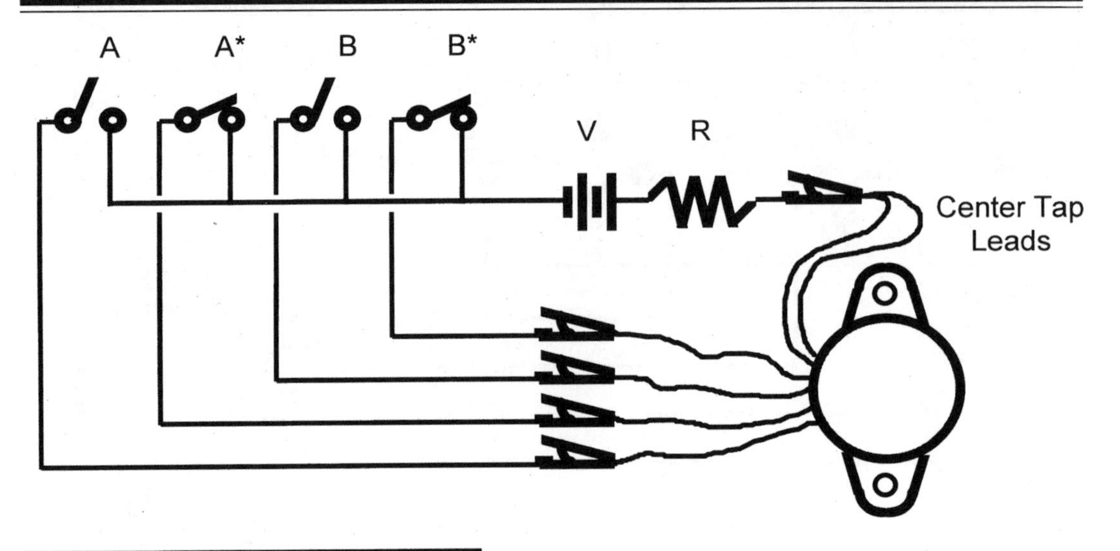

STEP	A	A*	B	B*
1	1	0	1	0
2	0	1	1	0
3	0	1	0	1
4	1	0	0	1
1	1	0	1	0

Figure 2-3. *Test setup and control sequence for a bifilar stepper motor.*

2. Arbitrarily connect the lead from switch A to one of the coil A leads. Connect the lead from switch A* to the other coil A lead.
3. Arbitrarily connect the lead from switch B to one of the coil B leads. Connect the lead from switch B* to the other coil B lead.
4. Tape a toothpick on the end of the stepper motor's rotor (like a compass needle, it will show the direction of movement).
5. Go through the control sequence of *Figure 2-1*. If the rotor does not go in the same direction for each step, reverse the coil A leads. Repeat and reverse the coil B leads if same direction motion is not observed. Repeat. If same direction motion is still not observed, reverse coil A leads once again.
6. Once same direction motion is obtained, reverse the control sequence and note the rotor moves in the opposite direction.

This test setup has the advantage in that it allows you to see the results of each control sequence step. Also, the batteries self limit the amount of current they can provide, protecting against any shorts. The test voltage of 3 volts for a 5 volt motor was selected because it is convenient (two D cells), and will provide enough power to energize the stepper motor (but will limit the amount of current to the motor). Even with these advantages, the setup is tedious and does not allow you to test for speed or torque. To do that, we need to move up in sophistication (and complexity).

Bifilar Test Setup

Figure 2-3 shows the test setup for a bifilar stepper motor, along with the control sequence.

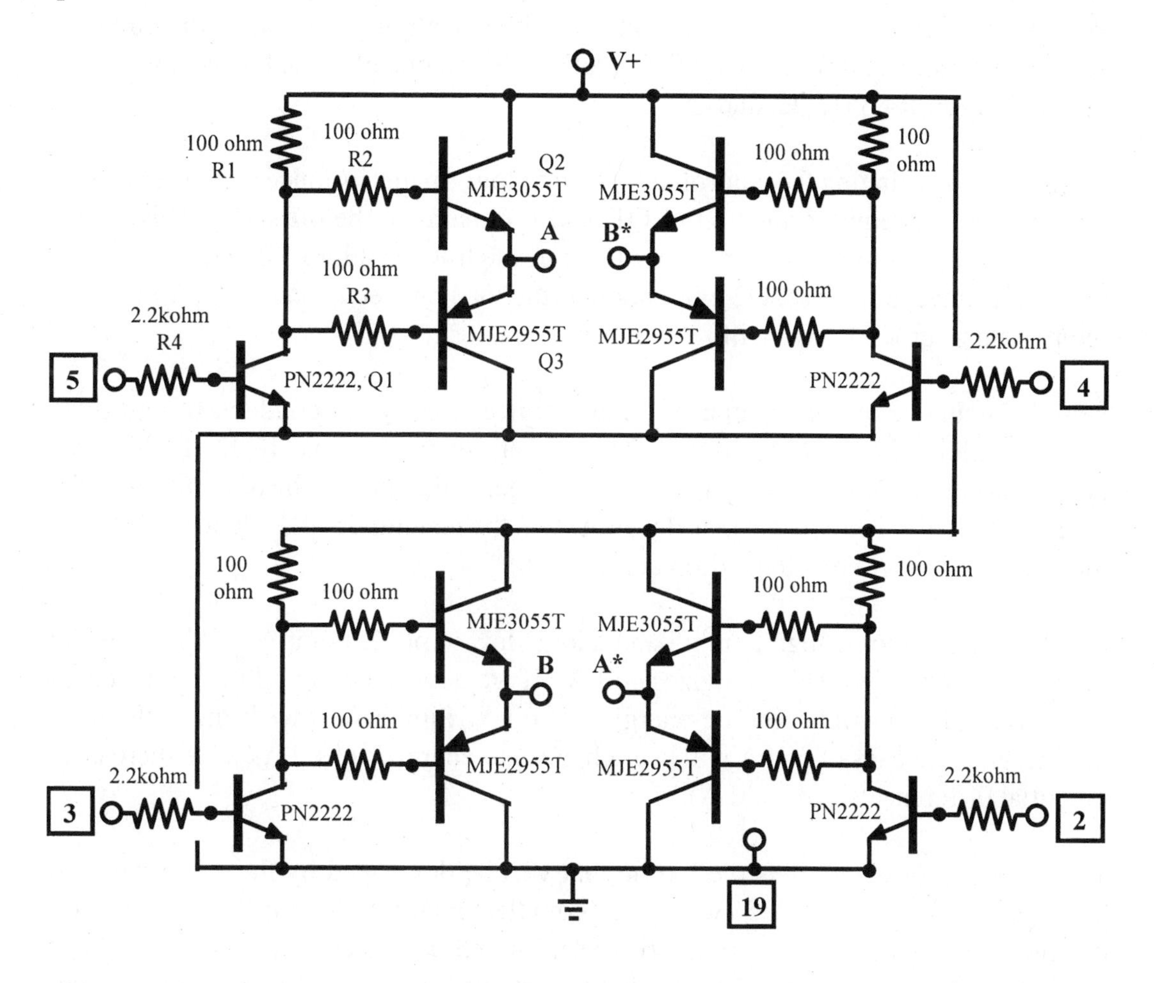

***Figure 2-4**. A control circuit.*

Control Electronics

To move from the test setup we just discussed to control electronics, we need to replace the four mechanical switches with electronic ones. Also, instead of physically moving the switches to change their status, we need a method to electronically control them from a relatively low power source (like a PC's parallel port). Such an electronic circuit is shown in *Figure 2-4*. Let's see how it works.

Control Input

The five boxed numbers refer to pins on the PC's parallel port. For pins 2, 3, 4, and 5, the PC provides either a logic 1 (high) or logic 0 (low) as referenced to ground (the PC's ground reference is provided to the circuit through pin 19). Since the circuit is actually four identical functional blocks, let's focus on one of those blocks, as defined by R1, R2, R3, R4, Q1, Q2, and Q3.

We'll begin by bringing the control input (pin 5) low. In this instance, no current flows through the base-emitter junction of Q1, and it remains in the off state. This allows current to flow from the power supply (V+) through R1 and into R2 and R3. With a positive voltage at its base, Q2 will conduct (the load between A and A* provides the return path to ground). Q3 will remain off.

Now let's change the pin 5 input to high. This causes Q1 to conduct, bringing the junction of R1, R2, and R3 to ground. The ground condition turns off Q2. It also turns on Q3, bringing point A close to ground. (Again, the current through Q3 and the resultant voltage drop across Q3 is dependent on the magnitude of the load. The larger the load, the larger the voltage drop across Q3).

So, if we apply a logic high to pin 5 and a logic low to pin 2, point A will be high and point A* will be low. This energizes coil A. If we now bring pin 5 low, both ends of coil A (A and A*) will be low, de-energizing coil A. Similarly, if we bring both pins 5 and 2 high, both ends of coil A will be high, also de-energizing coil A (since there is no potential difference across coil A).

The value of V+ determines the torque that will be developed by the stepper motor. The MJE3055 and MJE2955 (which are complementary NPN and PNP transistors) can provide about 1 amp of current. If we set V+ at 5.0 volts DC, there will be about a one volt drop across each transistor, leaving about 3 volts available for the stepper motor. We can increase V+ to about 6 volts, which will provide about 4 volts to the stepper motor. This increases the torque, still allows the TTL logic levels from a PC's parallel port to effectively control the PN2222 transistors, and keeps the current draw to a level that eliminates the need for heat sinks on the MJE transistors.

If you examine the control sequence for either coil A or coil B, you will see that one side is one while the other side is zero. So, an obvious question is, "Why not use a single parallel port pin, invert it in an additional 2N2222 transistor, and obtain the two required control signals?" It could be done, but it would not allow us to de-energize the motor. Remember, both voltages to the coil need to be either low or high to de-energize the coil. To do this, and provide the movement inputs, we need control over each individual control signal.

Control Software

The control software needs to provide the appropriate control sequence steps and delays between steps. If the control sequence steps are provided to the control electronics too quickly, the stepper motor will not be able to physically respond and will not rotate (in this instance, you would hear a low hum as the motor tries, but fails, to respond). The first thing the software needs to do is address the PC memory segment where the parallel port resides. (Please refer to my book, *Real-World Interfacing With Your PC*.) Then the software needs to establish the control sequence step values and send those values to the parallel port. Finally, the software needs to wait for a pre-established time before sending the next sequence step. A functional QBasic program for doing this is shown in *Program Listing 2-1*:

```
1  DEF SEG = 64 : P = 888

2  DIM A(4)

3  A(1) = 5 : A(2) = 12 : A(3) = 10 : A(4) = 3

4  DELAY! = 0.25

5:

6  OUT P, A(1) : GOSUB 13

7  OUT P, A(2) : GOSUB 13: REM: For reverse, A(4)

8  OUT P, A(3) : GOSUB 13: REM: For reverse, A(3)

9  OUT P, A(4) : GOSUB 13: REM: For reverse, A(2)
```

```
10 GOTO 5
11 END
12 ' SUBROUTINE DELAY
13:
14 START! = TIMER
15 WHILE (TIMER - START!) < DELAY
16 WEND
17 RETURN
```

Line 1 defines the proper memory segment to address the parallel port. It also sets the variable P to the standard address for a PC parallel port (888 decimal). Line 2 shows a four element numeric array which is filled in line 3. To see why those numbers were chosen, refer to *Figure 2-5*. The standard control sequence is shown in *Table A*. *Table B* shows the same control sequence, but with each "1" inverted to a "0," and each "0" inverted to a "1." This reflects the fact that the PN2222 transistors in the control electronics perform an inversion (a logic high input causes the transistor's collector-emitter junction to conduct, effectively bringing the collector to ground and creating a logic low). If we invert the control sequence number, it counteracts the inversion created by the control electronics and results in the proper control sequence. Finally, *Table C* rearranges the columns to place them in the numerical order of the parallel port pins. *Table C* also has an additional column ("Decimal") which lists the decimal equivalent of the binary bit pattern resulting from the associated logic highs and lows on pins 2, 3, 4 and 5 of the parallel port. Those decimal equivalents are the numbers used to populate the A array in line 3 of *Program Listing 2-1*.

Line 4 establishes the variable DELAY! (The exclamation point specifically defines this variable as a single precision decimal number, not an integer.) Actual motor control is performed in lines 6 through 9, where the control sequence steps are sent in turn to the control electronics through the parallel port. After each sequence step is outputted, the delay subroutine at lines 13 through 17 is performed. That subroutine gets the current value of the PC's internal timer and places it in the variable START! (line 14). Line 15 then loops until the timer's value has incremented an amount equal to DELAY!. When done, line 17 returns execution to the program statement that called the subroutine.

STEP	A	A*	B	B*
1	0	1	0	1
2	1	0	0	1
3	1	0	1	0
4	0	1	1	0
1	0	1	0	1

(A)

	PARALLEL PORT PIN			
STEP	5 (A)	2 (A*)	3 (B)	4 (B*)
1	0	1	0	1
2	1	0	0	1
3	1	0	1	0
4	0	1	1	0
1	0	1	0	1

(B)

	PARALLEL PORT PIN				
STEP	2	3	4	5	Decimal
1	1	0	1	0	5
2	0	0	1	1	12
3	0	1	0	1	10
4	1	1	0	0	3
1	1	0	1	0	5

(C)

***Figure 2-5**. Standard control sequences.*

Note the remarks in lines 7, 8 and 9. By changing the array reference to those shown in the remarks, the control sequence steps will be executed in reverse. This will cause the stepper motor to turn in the opposite direction.

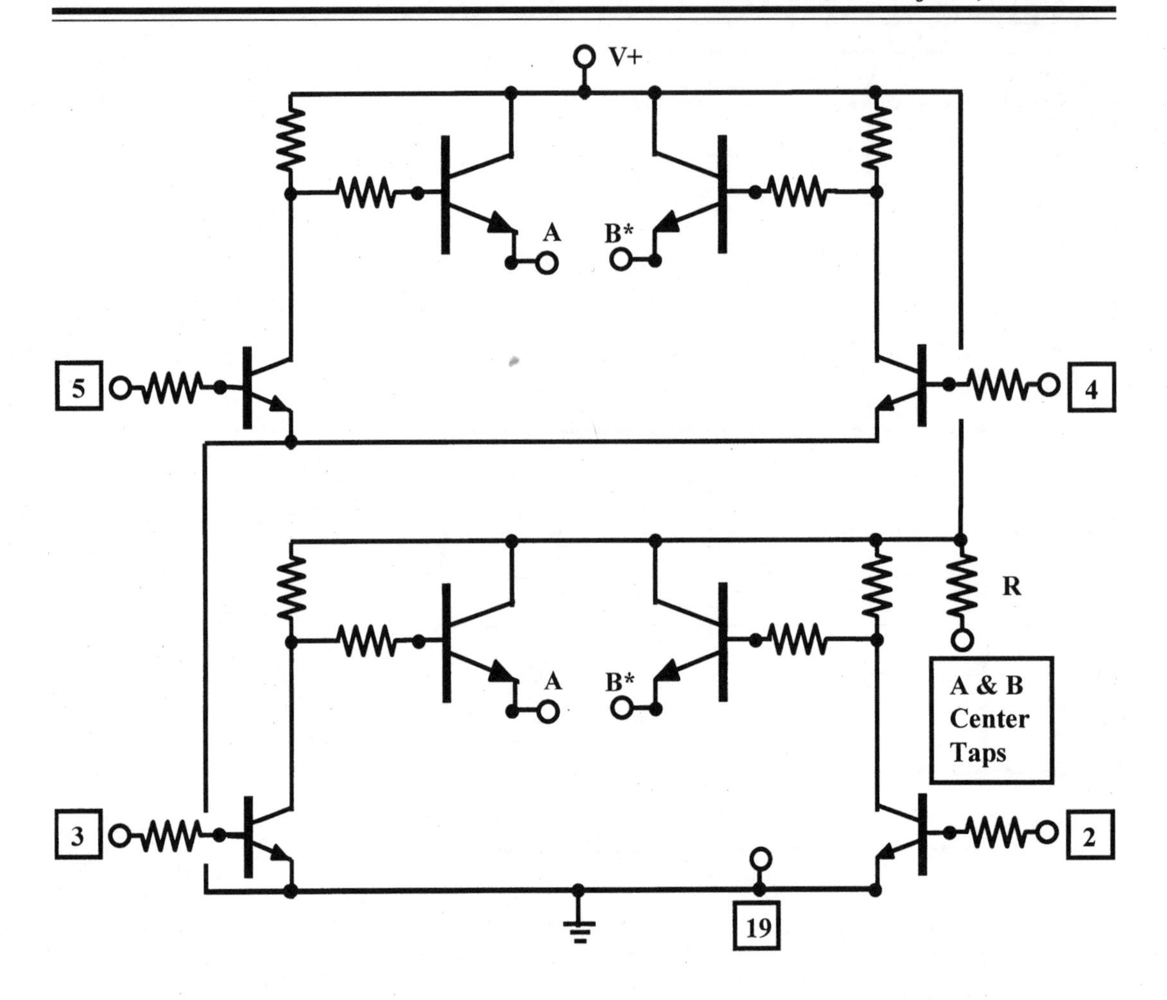

***Figure 2-6**. A control circuit for a bifilar stepper motor.*

Bifilar Control Electronics and Software

Figure 2-6 shows a control circuit for a bifilar stepper motor. Note that this circuit is similar to the bipolar control electronics; all component values are the same as the bipolar circuit. (See *Figure 2-4*) One main difference is the additional resistor R, which should be a 5 to 10 watt resistor between 10 and 20 ohms. The voltage (V+) should be approximately the same as the stepper motor's rating.

The control software shown in *Program Listing 2-1* can also be used for this application, providing the A array values are changed to A(1) = 10, A(2) = 9, A(3) = 5, and A(4) = 6 to reflect the bifilar control sequence.

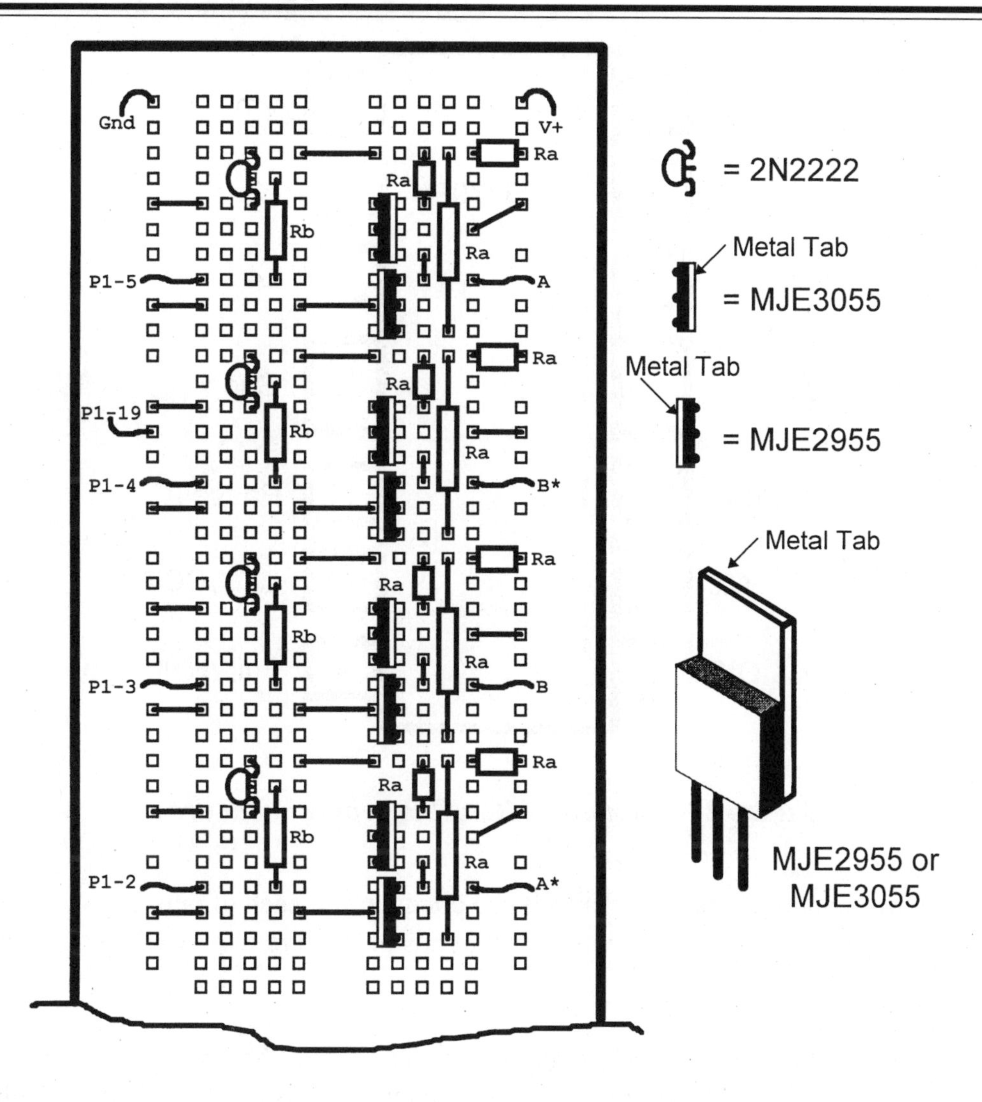

Figure 2-7. *How to construct the circuit of Figure 2-4 on a solderless breadboard.*

A Test Setup Using Control Electronics

The previous segment presented a test setup using switches. *Figure 2-7* shows how to construct the circuit of *Figure 2-4* on a solderless breadboard. The MJE3055 and MJE2955 transistors will not fit side-by-side without some adjustment. Install the first MJE3055 in the position shown. Then, push it towards the center of the solderless breadboard, bending the leads in the process. This will provide sufficient clearance to allow the MJE2955 to fit. Do this for all four MJE3055 transistors.

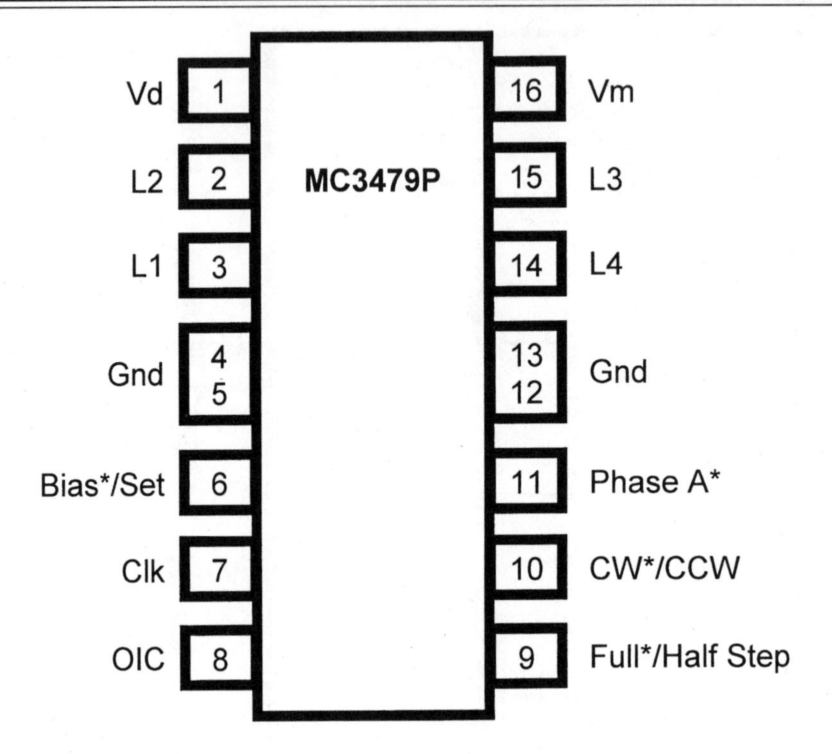

Figure 2-8. *Motorola MC3479P stepper motor driver.*

Make a five conductor cable and attach one conductor to each of pins 2, 3, 4, 5, and 19 on a DB-25 plug (male). Connect the other ends of the five wires to the points shown in *Figure 2-7.*

Attach a length of wire to each of the points marked A, A*, B, and B*. Install an alligator clip on the free end of each wire. Connect the positive lead from a 5 volt DC power supply to the point marked V+. Connect the negative power supply lead to the point marked *Gnd.*

Follow the procedure previously listed to determine which stepper motor wires to connect to points A, A*, B, and B*. Connect those wires. Type in *Program Listing 2-1*, connect the DB-25 plug to your PC's parallel port, power up the power supply, and run the program.

Of course, you can modify the circuit and the program as we've previously discussed, if you'd like to test bifilar stepper motors. All of the components used to create this test circuit can be reused when we begin our project.

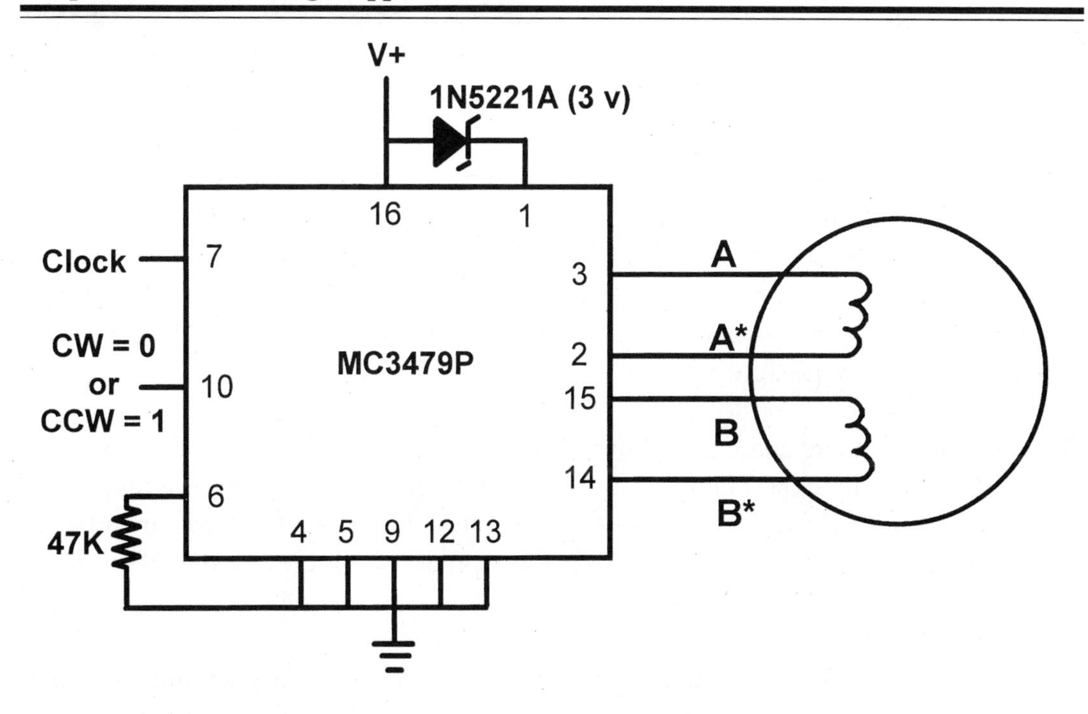

Figure 2-9*. A practical interconnection between an MC3479P and a stepper motor.*

Alternate Control Electronics

The previously described electronics use discrete transistors to control a stepper motor. An alternate approach is to use an IC specifically designed for stepper motor control. *Figure 2-8* shows such an IC, the Motorola MC3479P stepper motor driver. *Figure 2-9* shows a practical interconnection.

In use, pin 10 is made low for clockwise rotation (or high for counterclockwise rotation). A train of highs and lows are provided to pin 7, with each transition from low to high producing one step in the sequence. The power supply voltage, V+, should be about 5 volts plus the rating of the zener diode (as shown, it would require an 8 volt DC power supply).

The MC3479P can provide about 350 mA per coil. This is not sufficient for the stepper motors we'll be using (which require about 600 - 700 mA per coil). However, this IC can be used with smaller stepper motors.

The MC3479P is designed for use only with 2 phase bipolar stepper motors.

What's Next?

Now that you have a better understanding of how stepper motors work, and how to control them, you're ready to begin our main project. As you proceed through the chapters describing the mechanical and electronic construction, remember that the information contained in them has been structured to allow people with differing degrees of mechanical and electronic skills to complete construction successfully. You should consider applying your particular mechanical and/or electronic skills to improve or more easily construct the project.

For instance, if you are adept at making PC boards, you may elect to fabricate a board instead of using the experimenter's PC board referenced. As other examples, you might choose to fabricate some or most of the material using metal if you are adept at metalworking. You may choose to obtain a more precise lead screw in lieu of the 5/16" threaded rod specified.

As we proceed, you'll be provided information on critical design aspects and those that are subject to adjustment. (For instance, the table limit switch is shown situated on one of the table stepper motor supports. It can be located anywhere as long as the table trips it when it reaches a predetermined, repeatable position.) In addition, you'll also be provided information on critical requirements, so you can adjust the construction but maintain those requirements. (For instance, connection of the threaded rod and stepper motor shaft is shown by way of a section of shrink sleeving or poly tubing). If you're adept at metalworking, you could fabricate a metal coupling as long as the coupling keeps the shaft and rod concentric and true.)

We'll conclude with a detailed discussion of the software. This discussion should provide you with enough information to modify the software if you choose. If you choose to follow the book verbatim, you'll end up building a precise, useful, and practical project using stepper motors. In the process, you will also learn how to apply the skills you learn to your own future projects.

That's enough for now. Let's get on to the project!

CHAPTER 3

Getting the Parts Together

Introduction

The *Auto-XY* project isn't terribly complicated, but it does require a significant number of parts (including mechanical, electrical, and electronic). To avoid the frustration of diving into the project and then having to stop to buy this or that, this chapter lists all the parts by type (all electronic parts, all wood parts, and so on). With this information, you can gather all the parts *before* you begin construction. Also, to make part identification and construction more clear, all mechanical parts have been given a reference number. The reference numbers will be cited in the text and figures. If you have any doubt about exactly which part is being cited, refer back to the parts list using the reference number.

Electrical and Electronics Parts

Table 3-1 lists all the electrical and electronic parts for the project. The items with a -P suffix are for the power supply. Two of the items have critical requirements. If you choose to substitute another item for one listed in the table, you should first ensure yourself that the substitution meets the critical requirements listed next.

Stepper Motor Critical Requirement

The stepper motor is an Airpax model LB82773-M1. It is a 5 volt DC, 2 phase bipolar stepper motor, with a diameter of 2.25". Its body is 0.99" thick (not counting its 3/4" long, 1/4" diameter shaft) and weighs 0.61 pounds. The body has an integral mounting plate with 0.2" diameter mounting holes spaced 2.6" apart. Each of the motor's two field windings is rated at 5 volt DC, 800 mA based on a coil resistance of 6.25 ohms. The motor has a resolution of $7.5^{o}/o$ (48 steps per revolution), produces 100 g-cm of detent torque, and 1080 g-cm holding torque. Any other 2 phase bipolar 5 volt stepper motor can be used as long as its body diameter is not larger than $2^{1/4}$", it has a mounting plate with holes on about $2^{1/2}$" centers, it does not draw more than 800 mA per coil, and it has 100 g-cm or more of detent torque.

QTY	PART DESCRIPTION	CIRCUIT REFERENCE
1	1000 uF, 16v (min.) Radial Electrolytic Cap	C1-P
1	1 uf, 16v (min.) Radial Electrolytic Capacitor	C2-P
2	1N4003 Silicon Rectifier Diode	CR1-P, CR2-P
1	DPDT DIP 5VDC Relay, 50 ohm coil or greater.	K1
2	5vDC, 6.25ohm/coil, 2 Phase Bipolar Stepper Motor (Airpax LB82773-M1)	M1, M2.
1	DB-25 Connector (Male)	P1
4	MJE3055T NPN Transistor (TO-220 Case)	Q1, Q2, Q7, Q8
4	MJE2955T PNP Transistor (TO-220 Case)	Q3, Q4, Q9, Q10
6	PN2222 NPN Transistor (TO-92 Case)	Q5, Q6, Q11, Q12, Q13, Q14
1	Solderless Breadboard Type Printed Circuit Board	Controller Board
14	100 Ohm, 1/4w, fixed resistor	R1-R6, R9-R14, R18, R2-P
1	1kohm, 1/4 watt, PC mount Trim Pot	R1-P
9	2.2Kohm, 1/4w, fixed resistor	R7, R8, R15, R16, R17, R19-R22
2	"Micro" Switch	S1, S2
25 Ft	#22 Stranded Wire (length is approximate)	Hookup of various components
4 Ft.	12 Conductor Cable	Controller Board to Parallel Port
1	DB25 Male Connector with Hood	Controller Board to Parallel Port
1	Step down transformer (120vAC to 6.3vAC Center Tapped, 1.2A secondary)	T1-P
1	LM317T Voltage Regulator (TO-220 Case)	U1-P
1	Solderless Breadboard Type Printed Circuit Board	Power Supply Board
1	TO-220 Heat Sink	For LM317T
1	AC Duplex Outlet with Outlet Cover	Power for Drill
1	Normally Open Push Button Switch, 1A, 250V	Power for Drill
1	Small plastic project case	To house push button switch
1	6 foot AC Power Cord	Power Supply AC
1	250v, 1 Amp Fuse and In-Line Fuse Holder	Power Supply AC

Table 3-1*. A list of all the electrical and electronic parts for the Auto-XY.*

Relay Critical Requirement

The double pole, double throw (DPDT) relay is an Aromat DS2M5 PC-mount type in a 16 pin dual in-line package (DIP) configuration. It has a 5 volt DC, 62.5 ohm coil, although any similar device with a coil resistance of 50 ohms or greater can be driven by the circuitry. The relay's pinout is shown in *Figure 3-1.* If a relay with a different pinout is used, the electronic construction details will have to be modified to accommodate the different pinout.

Solderless Breadboard Type Printed Circuit Board

Two PC boards are required, one for the controller electronics and one for the power supply electronics. These are printed circuit boards that have the same configuration as a standard solderless breadboard. They are available at most nationwide electronics retail outlets. As an example, Radio Shack carries such a board as part number 276-170.

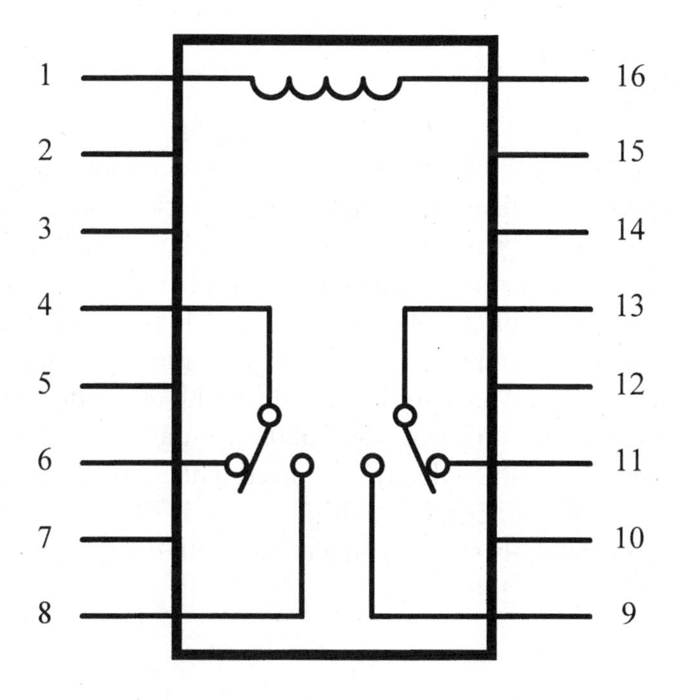

Figure 3-1*. The pinout for the relay.*

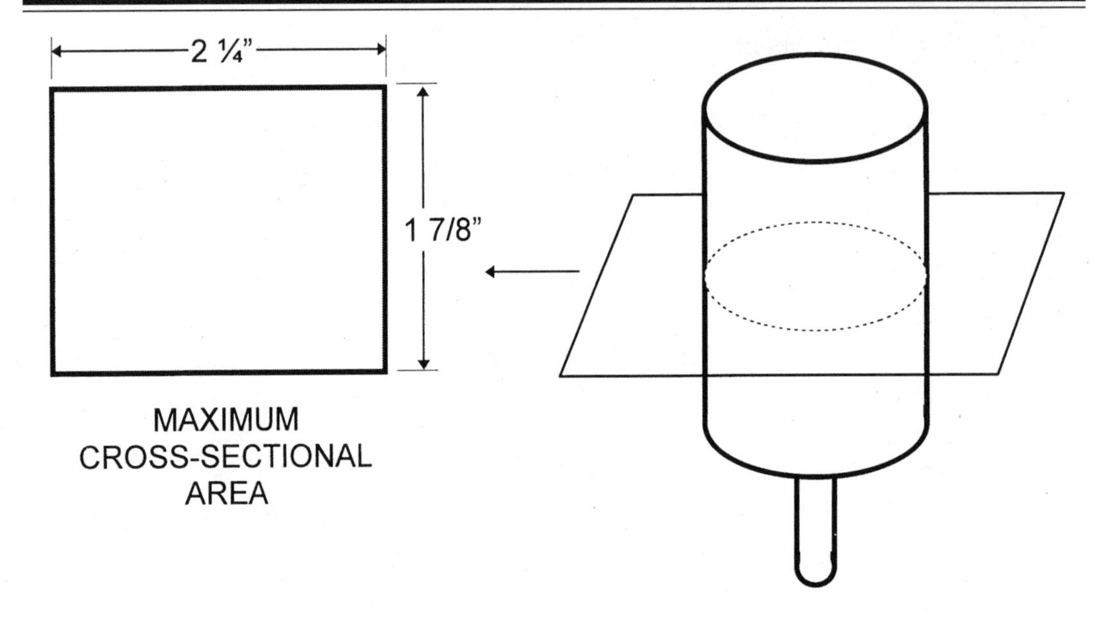

Figure 3-2. *Cross sectional dimension requirements for the Auto-XY.*

The *Auto-XY* Drill

The *Auto-XY* project uses a mini-drill (sometimes called a moto-tool or hobby drill). This is a small, lightweight, hand-held, high speed electric drill. Since the mini-drill must be precisely moved and positioned by the drill caddy stepper motor, and the size of the drill caddy is limited by other factors, the mini-drill is a critical electromechanical component. Critical requirements are: one pound or less net weight, and cross sectional dimensions as shown in *Figure 3-2*. (If the larger cross-sectional dimension is greater than $2^{1/4}$", you will have to increase the width of the drill bridge as defined in Chapter 7.) One possible drill is model 2620, manufactured by S-Turbo D.I.Y. & Hardware Company, LTD., although any drill meeting the critical requirements can be used. Since the mini-drill will be used to drill holes in PCBs, a #64 (.036"D) PC drill bit with a 1/8" shaft should also be obtained at the same time as you obtain the mini-drill.

Hardware

Hardware is used throughout the *Auto-XY* project to fasten mechanical parts, drive the table and drill assemblies, and ensure smooth operation. All hardware required is listed in *Table 3-2*.

#	QTY	DESCRIPTION	FOR	NOTE
1	28	#4 x 1/2" Sheet Metal Screw	Table/Drill Guides, Case Rear & Bottom	
2	18	#8 x 3/4" Sheet Metal Screw	Drill Caddy Support, Drill Caddy, Table Mount to Case	
3	19	#4-40 x 3/4" Machine Screw	Various Mounting	
4	12	#4-40 x 1" Machine Screw	PC Board, Table Fence Mounting	
5	35	#4-40 Nut	For #4-40 Screws	
6	8	#6 Flat Washer	Drill Bridge and Table Stepper Mount Mounting	
7	1	#6-32 x ¾" Machine Screw	AC Outlet Mount	
8	6	#6-32 x 1 ¼" Machine Screw	Table Stepper & Drill Bridge Mounting	
9	4	#6-32 x 2" Machine Screw	Drill Stepper & Table Stepper Mount Mounting	
10	2	#8-32 x 4" Machine Screw	Drill Collar	
11	14	#6-32 Nut	For #6-32 Screws	
12	4	#8-32 x 1 1/4" Flat Head Machine Screw	Table Tracks to Base	
13	8	#8 Flat Washer	Table Tracks to Base	
14	14	#8-32 Nut	For #8-32 Screws	
15	2	¼" x 4" Lag Bolt (3" unthreaded shaft)	Drill Guide	
16	4	¼" Nut and appropriately sized washers	for Lag Bolt	
17	2	5/16"-18 or 5/16"-24 Threaded Rod, 12" long	Table & Drill Drive	
18	2	5/16" Nut to match Threaded Rod	Table/Drill Drivers	
19	2	Cable Clips	Controller, AC Cable	
20	2	1" long, ¼" ID Poly Tubing		1
21	2	1" long, ¼" diameter Heat Shrink Tubing		1
22	1	3/8" x ¾" x .032" Utility Compression Spring	Drill Caddy	

***Table 3-2**. Hardware list for the Auto-XY.*

A Special Note on Mechanical Linkages

The stepper motors' shafts need to be mechanically linked to the 12" long, 5/16" threaded rods (item 17) in a precise manner. That is, the shaft and rod need to be as close to perfectly concentric as possible. This can be accomplished in various ways. One way is to obtain a 1" section of 1/4" inside diameter poly tubing and a tubing clamp. (3/8" outside diameter, 1/4" inside diameter poly tubing should be available at most national home centers.)

As an alternate to the poly tubing, you can use 1/4" diameter heat shrink tubing. One end of a 1" length of tubing is expanded slightly by pushing in the end of a standard ball point pen into one end of the tubing. The expanded end is "screwed" on one end of the threaded rod to a distance of about 1/4". At the appropriate point in the construction process, the other end is passed over the shaft of the stepper motor and the tubing is shrunk onto both the rod and shaft by using a heat gun or a heavy duty hair dryer.

A Special Note on the Threaded Rod

The threaded rod specified item (17) is a 5/16" type found at most hardware stores. The threads per inch might not be specified. Instead the rod may be specified as "USS Coarse". You may choose to use a different size threaded rod (like 1/4") with a different number of threads per inch. During the calibration, we'll describe how to adjust the software's setup file to accommodate any size threaded rod.

A Special Note on the Lag Bolt

The 1/4" x 4" lag bolt (item 15) will have its head cut off. The short (about 1") threaded portion will affix the bolt onto the drill caddy. The unthreaded portion (about 3") will serve as one of the two drill collar guides. It will also receive a spring. When fully assembled, the drill collar will ride on the unthreaded portions of two lag bolts, with spring tension provided by two springs. This will allow the drill to move precisely up and down, without any lateral motion. For this reason, the lag bolts should be a straight as possible.

A Special Note on the Compression Spring

The compression spring (item 22) is specified as 3/8" diameter by 3/4" long (or longer up to 1 1/2" long), constructed of 0.32" spring wire. This particular size may not be

#	QTY	DESCRIPTION	FOR	NOTES
32	2	Tube Clamps (#2, 7/32" - 5/8" adjust)		1
33	2	1/8" Thick Flat Aluminum Stock, 4 ½" long	Drill Caddy Side Stretcher	2
34	2	Angle Aluminum, 2 1/4" long	Table/Drill Guide	
35	4	Angle Aluminum, 3" long	Drill Bridge Verticals	
36	1	Angle Aluminum, 5 3/4" long	Drill Motor Mount Track	2
37	1	Angle Aluminum, 9 3/4" long	Short Drill Caddy Track	
38	1	Angle Aluminum, 10 ½" long	Long Drill Caddy Track	
39	1	Angle Aluminum, 14" long	Long Table Track	
40	1	Angle Aluminum, 12" long	Short Table Track	

***Table 3-3**. These items need to be fabricated from aluminum stock.*

readily available at your location. Any compression spring constructed from 0.32" (or thinner) spring wire that will easily fit over the 1/4" lag bolt (item 15) will do. The spring should allow at least a 1/2" "throw" between the fully relaxed and fully compressed positions. Most national home centers carry a spring assortment from spring manufacturer Century Spring Corporation, Los Angeles, CA.

Aluminum Stock

The items to be fabricated from aluminum stock are listed in *Table 3-3*. Two types of aluminum stock are used. They are flat aluminum bar stock, 3/4" wide by 1/8" thick, and angle stock, 3/4" x 3/4" x 1/8" thick. The bar and angle stock should be readily available at most national home centers.

Wood

The items to be fabricated from wood are listed in *Table 3-4*. Various parts of the project are fabricated from dimensional lumber and plywood. The dimensional lumber can be readily available 3/4" thick pine, or other commonly available woods like poplar and mahogany. The plywood is 1/4" thick Luan mahogany. In addition, pegboard (commonly called "masonite") is used for the case bottom and rear. The holes in the pegboard allow for natural air flow, ensuring the controller board transistors receive a sufficient flow of cooling air when housed in the case.

#	QTY	DESCRIPTION	FOR	NOTES
41	1	8" x 10" x ¼" plywood	Table Surface	
42	2	1" x 1 5/8" x 3/4" Thick Wood	Table Stepper Mount	
43	1	1" x 2 1/4" x 3/4" dimensional lumber	Table Driver	
44	1	14" x 19" x ¼" plywood	Base	
45	3	5 ½" x ¾" x ¾" dimensional lumber	Small Drill Leg Spreader	2
46	1	5 ½" x 1 ½" x ¾" dimensional lumber	Large Drill Leg Spreader	2
47	2	3 ¼" x 1 ¾" x ¾" dimensional lumber	Drill Holder	2
48	1	1 ½" x 2" x ¾" dimensional lumber	Drill Driver	2
49	1	1" x 2" x ¾" dimensional lumber	Drill Caddy End Piece	2
50	4	¾" x 19" x ¾" dimensional lumber	Case Front & Back Rails	
51	4	¾" x 2" x ¾" dimensional lumber	Case Front & Back Stiles	
52	2	3 ½" x 14 ¾" x ¾" dimensional lumber	Case Sides	
53	1	14" x 19" x 1/8" pegboard	Case Bottom	
54	1	3" x 18 ¼" x 1/8" pegboard	Case Rear covering	
55	1	2 ½" x 18" x ¾" dimensional lumber	Case Drawer Front	
56	1	1 ¾" x 17" x ¼" plywood	Case Drawer Rear	
57	1	8 ¼" x 17" x ¼" plywood	Case Drawer Bottom	
58	2	1 ¾" x 8" x ¼" plywood	Case Drawer Sides	
59	A/R	Carpenter's Glue	Wood Assembly	
60	A/R	Miscellaneous pieces of dimensional lumber (including small amounts of hardwood, like mahogany or walnut)	Miscellaneous	2

***Table 3-4**. These items need to be fabricated from wood.*

Tools and Techniques

As you have reviewed the different tables and text in this chapter, you may have already realized that you will need access to (and appropriate skills to effectively use) various tools. Although every effort has been made to minimize the need for specialized tools and techniques, some are required to ensure a precision result.

Bench Saw

A few wood pieces need to be cut precisely and perfectly square. The most appropriate tool for this task is a stationary bench saw with a good miter gauge and fence. If you do not have access to a bench saw, the few wood pieces needing this precision can be cut by a friend or local carpenter's shop. If you are adept at metalworking, you can substitute metal pieces for these few wood pieces.

Drill Press

Some drilling also needs to be performed precisely. The most appropriate tool for this task is a radial arm drill press. A drill press stand for a hand-held drill will also work adequately, providing you take the necessary time to set it up precisely. In addition, you'll need various drill bits, including 3/32", 1/8", 5/32", 3/16", 7/32", 1/4", 9/32", 5/16", 3/8", and 1/2" diameters.

Other Tools

In addition, you will need other hand tools:

1. Variable speed, 3/8" hand-held drill.
2. Hacksaw or other metal cutting saw.
3. Flat blade screwdrivers, adjustable and open end wrenches, wood planes, assorted clamps, etc.
4. Assorted files and sandpaper.

CHAPTER 4
Building the Case and Base

Introduction

The case is the assembly that has the *Auto-XY* base as its top. It serves as a "foundation" for the *Auto-XY*, and also includes a drawer to hold drill bits and various other items that you will use with the *Auto-XY*. In addition, the case houses the control electronics and power supply.

The base (#44, in *Table 3-4*) is the plywood platform on which the table stepper, guides, and drill caddy are mounted. This chapter describes the construction of these two items.

The Case

Begin construction by cutting the case front and back rails (#50), case front and back stiles (#51), case sides (#52), and case drawer front (#55) from 3/4" thick dimensional lumber (which we'll refer to as just plain "wood" from now on).

Case Front and Back

You will now construct two identical elements; one which will serve as the case front, and the other the case back. To make the case front, assemble two case rails (#50) and two case stiles (#51) as shown in *Figure 4-1*, after first coating all mating surfaces with carpenter's glue (#59). Align all mating surfaces accurately, and then clamp together, applying some (but not too much) pressure on the joints. Let the joints set for about an

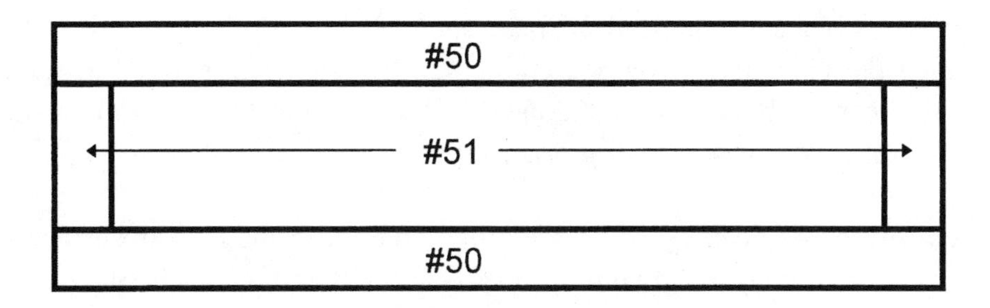

Figure 4-1. *Assemble the case rails and stiles as shown.*

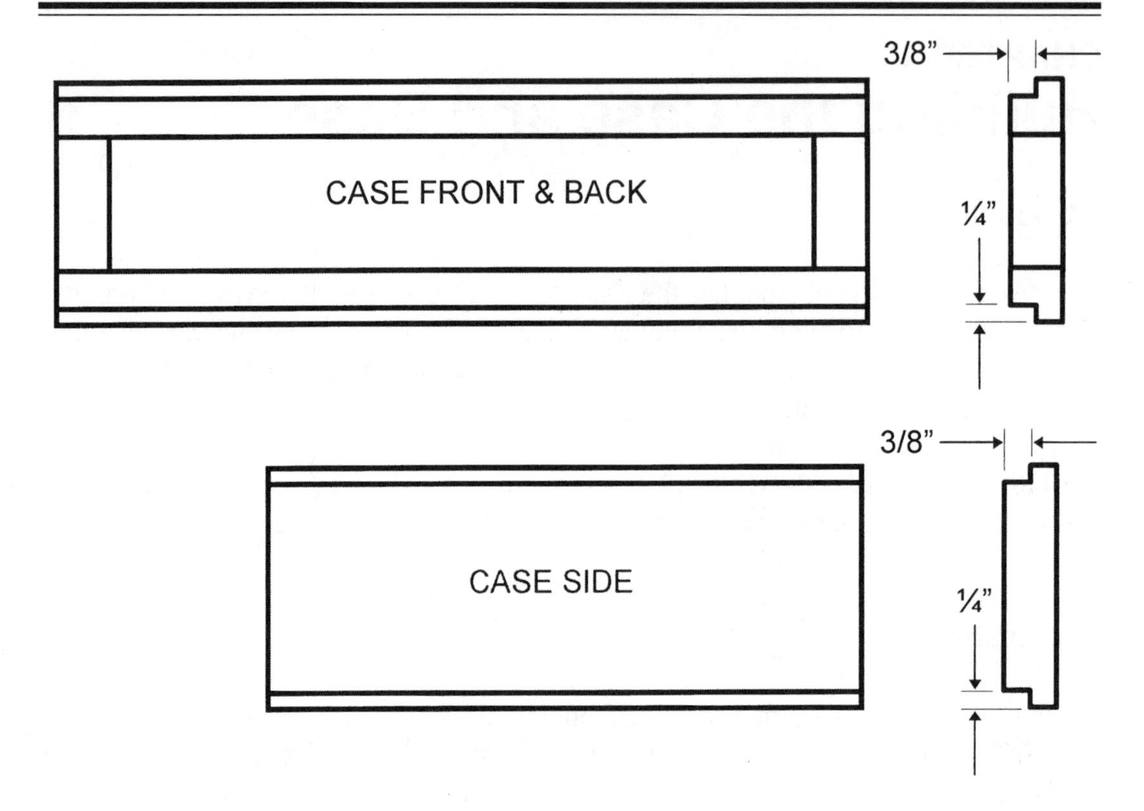

Figure 4-2. *Cut a rabbet in the top and bottom of the case front, sides, and rear as shown.*

hour before removing the clamps. Sand all surfaces smooth. Repeat this process to fabricate the case back.

Forming the Rabbets

The case front, sides, and rear need to have rabbets cut along the length of their top and bottom surfaces (to accept the *Auto-XY* base and case bottom, respectively). In addition, the case sides need to have rabbets cut in their front and rear surfaces to accept the case front and rear. Rabbeting is best done on a bench saw, although good results can be obtained with a straight-cutting bit in a router.

Begin by cutting a 1/4" deep by 3/8" wide rabbet in the top and bottom surfaces of the case front, sides, and rear as shown in *Figure 4-2*. Next, cut a 3/4" deep by 3/8" wide rabbet in both sides of both case sides as shown in *Figure 4-3*. These rabbets will accept the sides of the case front and back.

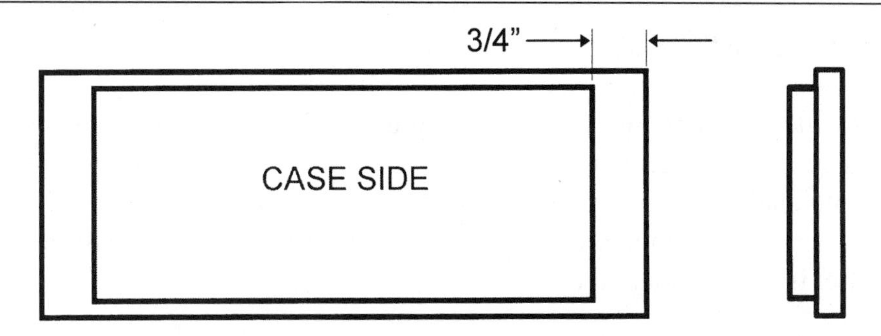

Figure 4-3. *Cut a rabbet in both sides of both case sides, as shown.*

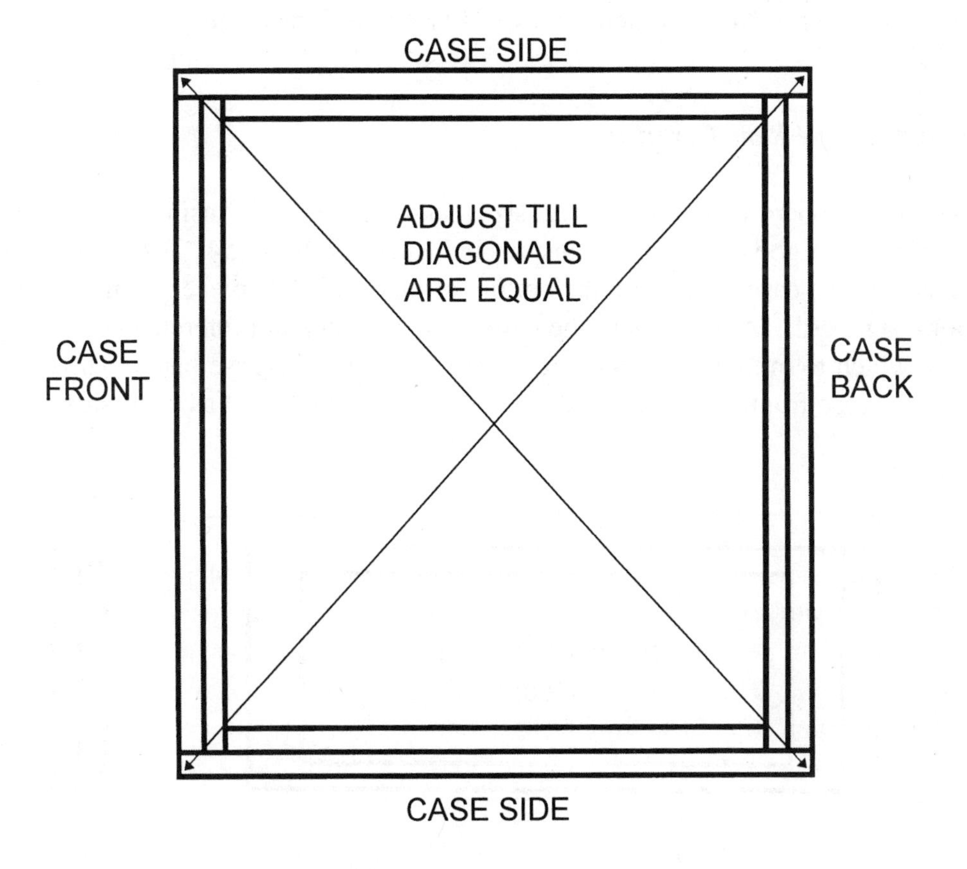

Figure 4-4. *Measure the two diagonals as shown.*

Assembling the Case "Carcass"

Coat the 3/4" deep rabbets with carpenter's glue (#59) and assemble the case front, sides, and back to create a rectangular box as shown in the *Figure 4-4* top view. Secure the pieces with clamps. Measure the two diagonals (as shown in *Figure 4-4*) and push opposing corners as appropriate until both diagonals are equal. This ensures a square assembly. Let it sit for about an hour.

Adding the Case Bottom

Cut a rectangular piece of 1/8" thick pegboard to the dimensions indicated for item #53. Test fit this piece into the rabbets on the bottom of the case. (Either surface may be arbitrarily identified as the bottom, with the other surface becoming the top.) Adjust the dimensions of the pegboard as necessary to fit into the case bottom rabbets. Secure with twenty #4 sheet metal screws (#1) equally spaced around the perimeter of the pegboard.

Adding the Case Rear Covering

A piece of pegboard will now be cut to size and installed on the inside of the case back. Cut a piece of pegboard to the dimensions indicated for item #54. Test fit the piece on the inside of the case back, with the bottom of this pegboard piece resting on the case bottom pegboard. (At this point, the case front and back are interchangeable. Select one on which to mount this pegboard — this becomes the case back. The other becomes the case front.) Adjust the dimensions of the pegboard as necessary to fit the

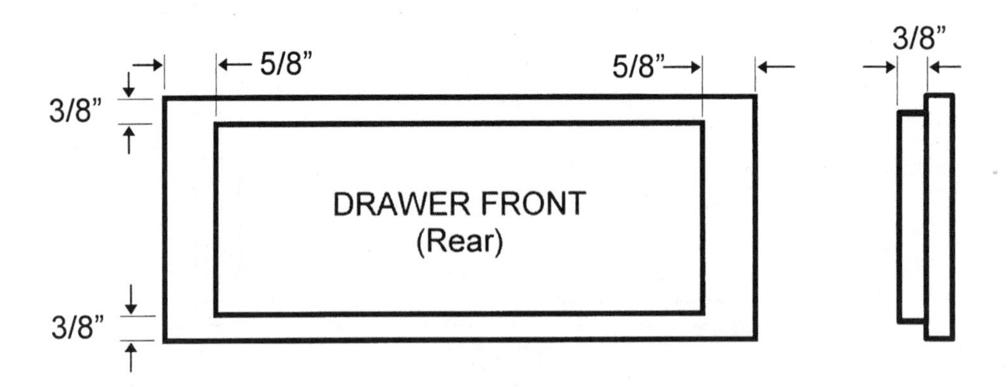

***Figure 4-5**. Form the rabbets in the drawer front as shown.*

case back, ensuring the pegboard does not extend into the upper rabbet area. Secure with three #4 sheet metal screws (#1) spaced equally along each long horizontal side, and one screw in the center of each short vertical side (for a total of eight #4 sheet metal screws).

Building the Case Drawer

The case drawer consists of the drawer front (#55), drawer rear (#56), drawer bottom (#57), and two drawer sides (#58). Cut all of these items to the dimensions indicated in *Table 3-4*.

Forming the Drawer Front

The drawer front will have rabbets on all four sides. The bottom rabbet will accept the drawer bottom. The side rabbets will accept the drawer sides. The top rabbet will let the complete drawer assembly to fit inside the case front cavity, allowing the drawer front to appear only 1/4" thick. After cutting the wood to size, form the rabbets in the drawer front (#55) as shown in *Figure 4-5*.

Dry Fitting the Case Sides, Bottom and Rear

Using some masking (or other available) tape, fit the various components of the case drawer together to ensure they fit properly. Then, slide the dry-fitted drawer into the case front to ensure that it fits. Make any adjustments to the components' dimensions at this point to allow the drawer to fit together properly, and to fit into the case. Remove the tape from all parts.

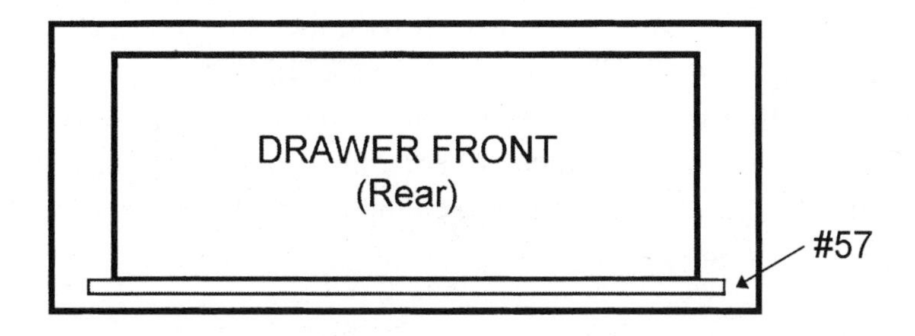

Figure 4-6. *Place the drawer bottom into the glued rabbet and center as shown.*

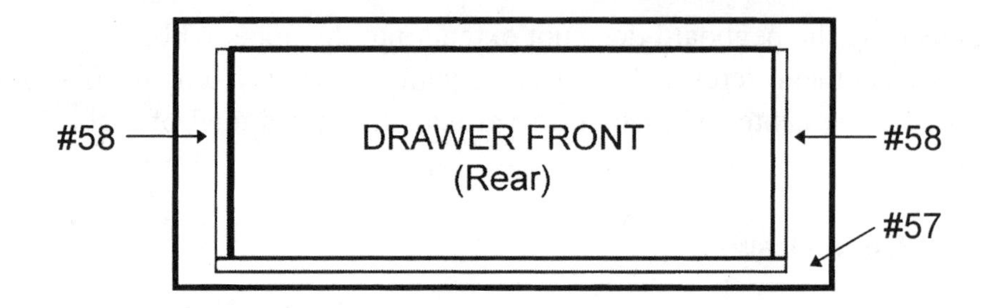

Figure 4-7. *Place each drawer side into one glued rabbet, and move it flush with the outside edge of the drawer bottom as shown.*

Attaching the Drawer Bottom to the Front

Apply carpenter's glue (#59) to the drawer front's bottom rabbet. (Select one of the longer sides to be the drawer front bottom. The other long side then becomes the drawer front top.) Place the drawer bottom into the glued rabbet, centering it along the rabbet's length, as shown in *Figure 4-6.* Clamp it into place and allow it to set for about 30 minutes. If you have access to a heavy duty staple gun with 1/2" staples, you can secure the plywood pieces to each other and the drawer front with these staples (in addition to the glue). If you do this, you will not have to clamp the pieces together and wait until the joints cure.

Attaching the Drawer Sides

Apply carpenter's glue (#59) to the drawer front's side rabbets and along the bottom edge of each drawer side (#58). (Select one of the longer edges to be the drawer side bottom. The other long edge then becomes the drawer side top.) Place each drawer side into one glued rabbet and move it flush with the outside edge of the drawer bottom, as shown in *Figure 4-7.* Clamp into place and allow to set for about 30 minutes. If you have access to a heavy duty staple gun with 1/2" staples, you can secure the plywood pieces to each other and the drawer front with these staples (in addition to the glue). If you do this, you will not have to clamp the pieces together and wait until the joints cure.

Attaching the Drawer Rear

Apply carpenter's glue (#59) to the drawer side and bottom exposed edges. Position the drawer rear (#56) against the glued edges as shown in *Figure 4-8.* Clamp into place and allow to set for about 30 minutes. Again, use a heavy duty staple gun with 1/2" staples to secure the plywood pieces to each other if you do not want to clamp the pieces together and wait until the joints cure.

Adding Drawer Spacers to the Case

If you now put the drawer in the case, it will tip toward the case bottom and wobble from side to side. To prevent this, side and bottom spacers need to be fabricated and installed in the case. *Figure 4-9* shows a cutaway view of the case, looking from the rear towards the front. Note the two side and two bottom spacers.

From miscellaneous dimensional lumber (#60), cut four pieces of wood to a thickness of 3/4" and a length of 8". Measure the distance between the case's pegboard bottom (#53) and the bottom of the case's drawer opening. Cut two of the four pieces of wood to that height. Secure these two pieces to the pegboard bottom with carpenter's glue (#59).

Install the drawer in the case and measure the distance between the drawer side (#58) and the case side (#52). Cut the two remaining wood pieces to that dimension. Temporarily remove the drawer and tape the two wood pieces into the positions shown in *Figure 4-9*. Reinsert the drawer and see if it will fit without binding. If the wood pieces are too large, either sand or plane them down to allow the drawer to slide in and out without binding. When you've obtained a good fit, glue the wood pieces to the case sides in the positions shown in *Figure 4-9* with carpenter's glue (#59). Tape the pieces in place and leave undisturbed. After about 30 minutes, remove the tape and reinstall the drawer.

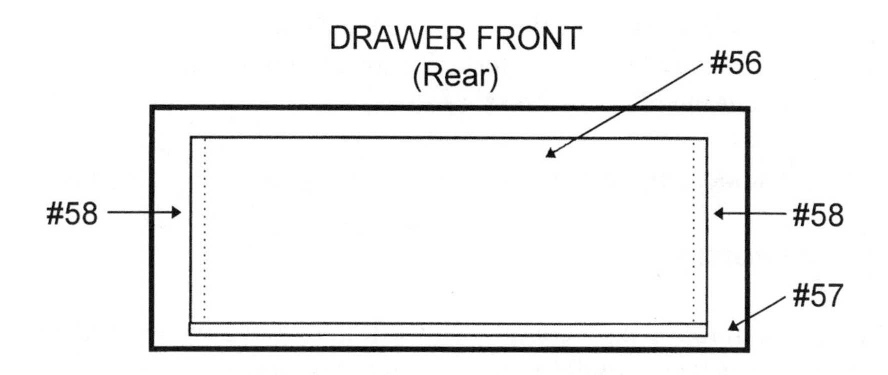

Figure 4-8. *Position the drawer rear against the glued edges as shown.*

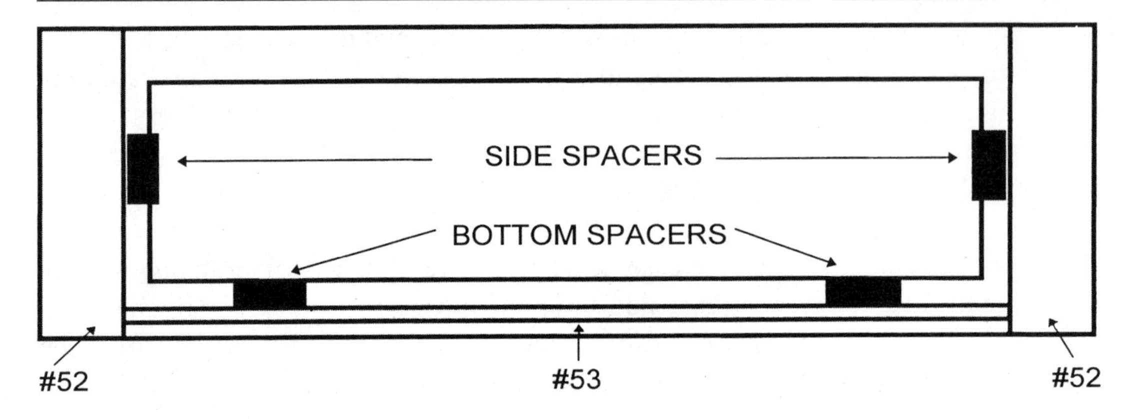

Figure 4-9. *A cutaway view of the case, looking from the rear toward the front.*

Fabricating the *Auto-XY* Base

The *Auto-XY* base (#44) fits into the rabbets on the top of the case. It is the base for the table, drill caddy, and all other associated items. Begin fabricating the base by cutting a piece of 1/4" plywood to the dimensions indicated for item #44. Test fit item #44 into the rabbets in the case. Adjust the dimensions as necessary until item #44 fits into the case rabbets.

Figure 4-10 contains the locations for the 10 required 1/4" diameter round (and one rectangular) holes. Choose one corner of item #44 as the zero reference (lower left). From that zero reference, lay out the positions of the 10 round holes. Drill a small pilot hole at each location, and then enlarge each hole to 1/4" diameter.

Draw the outline of the rectangular hole using the dimensions shown in *Figure 4-10*. The hole can be cut out using a sabre saw, after drilling a 3/8" hole in each corner (within the rectangular hole's perimeter). To avoid splintering of the wood, use a plywood or metal cutting blade in the sabre saw.

Sand the surface smooth and cover with a clear coating of water based varnish.

On to the Electronics

With the case and *Auto-XY* base completed, we'll now move on to building the electronics. The completed electronics will allow us to test and align the table and drill as we construct them. Once we're completely done with construction, we'll install the electronics into the case.

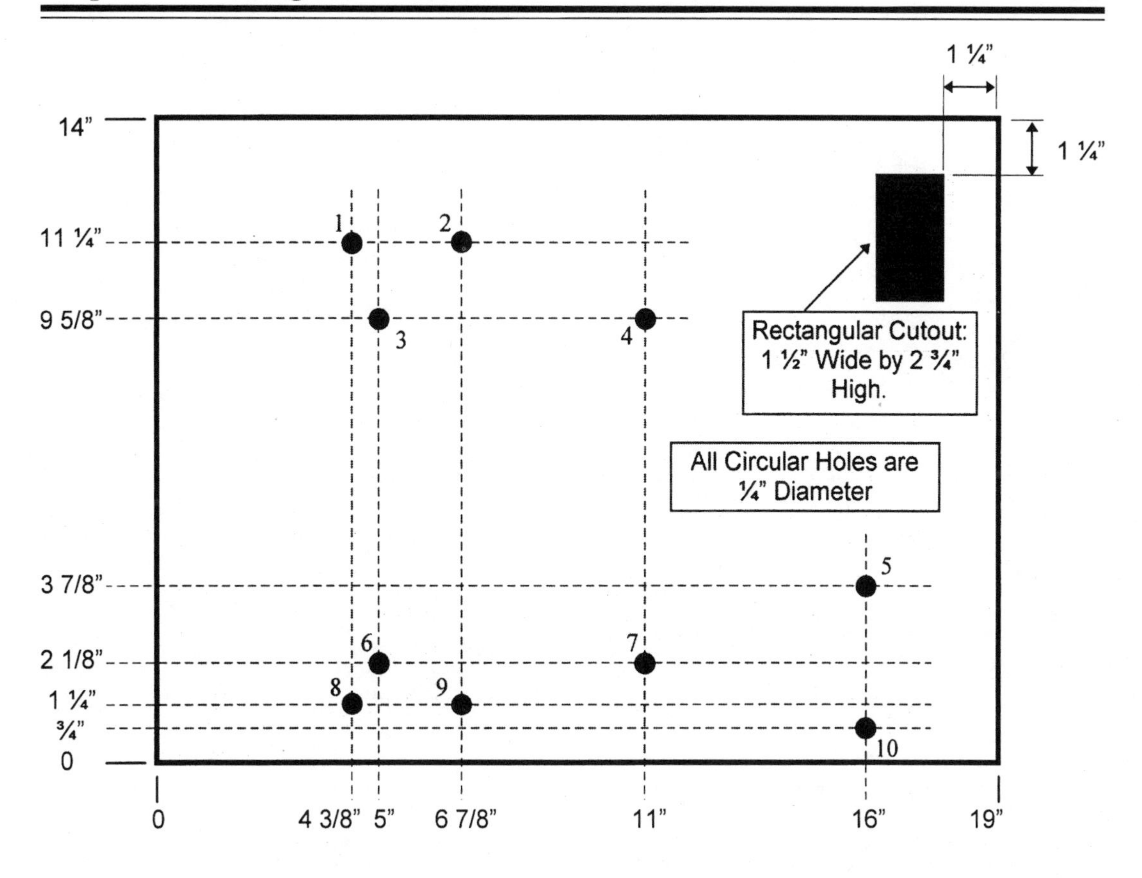

***Figure 4-10**. Locations for the holes.*

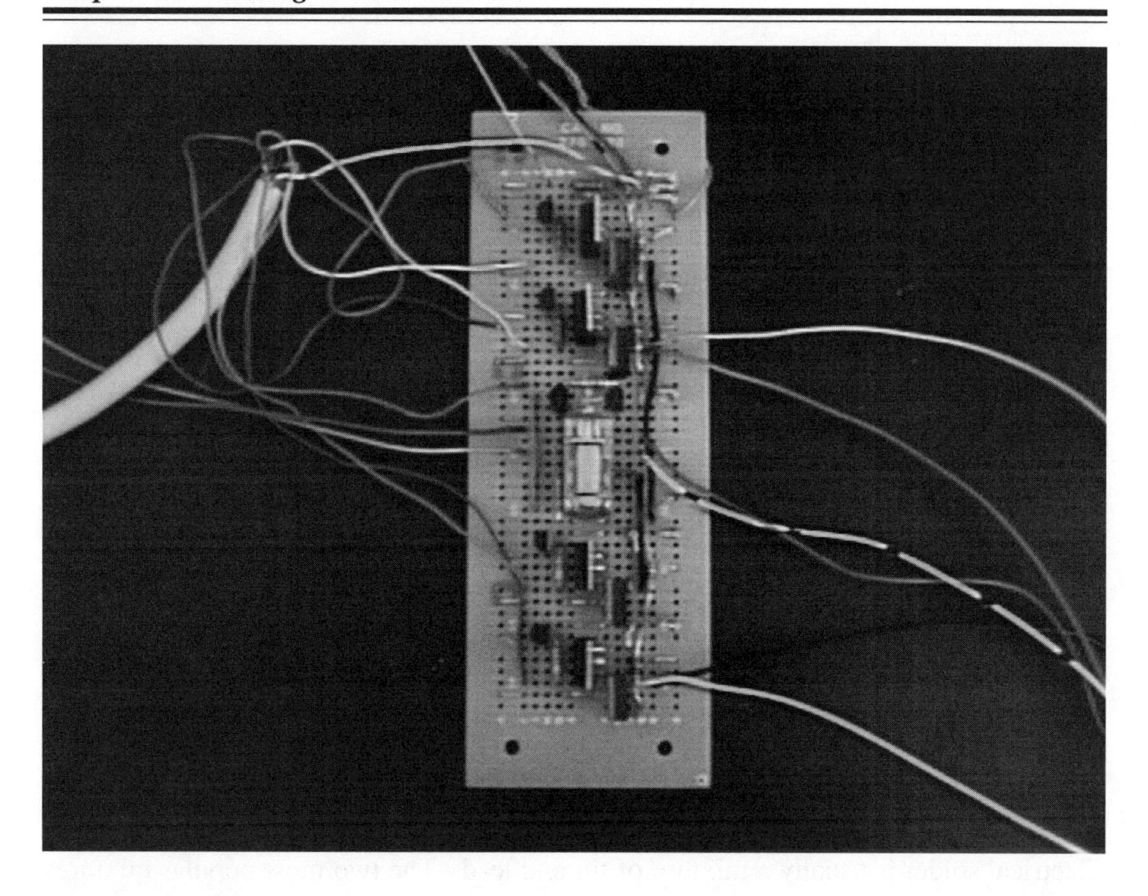

CHAPTER 5
Building the Electronics

Introduction

The *Auto-XY* electronics consist of the controller PCB and the power supply PCB. The controller PCB contains the interface between a PC's parallel port and the stepper motors (and their associated limit switches). The power supply PCB accepts unregulated AC voltage from the step down transformer, converts it into DC voltage and then regulates it for use on the controller PCB.

Putting together these electronics at this stage of construction will allow you to test out your stepper motors and limit switches. As you proceed to build the remainder of the project, the electronics will allow you to test and adjust the table and drill assemblies.

Construction Techniques

Construction consists of populating the board with jumpers, components, and wire leads. Step-by-step instructions are provided in the tables, and the results are reflected in the like-numbered figures. It is a good idea to check off each step as you proceed, and compare the table instruction, associated figure, and your PCB after each step is completed. However, before beginning, let's take a few minutes to refresh ourselves on proper soldering, lead forming, component handling, and wire preparation techniques.

The Solder Joint

To create a good joint, all surfaces must be in actual contact. The surfaces must also be free of oxidation (since oxidation forms a layer above the metal and thus prevents the formation of a mechanically sound solder joint). The surfaces to be joined and the solder must be heated to the point where the solder will be in a liquid state. When the heat is removed, the joint must remain motionless while the solder again returns to a solid state.

The Right Solder

Electrical solder is usually a mixture of tin and lead. The two most popular mixtures are SN63 (which is 63% tin and 37% lead), and SN60 (60% tin and 40% lead). The third element in the solder is flux, usually rosin, imbedded in the core of the solder (hence the name "rosin core" solder). The flux is very important, because it removes (and keeps from reforming) any oxides on the parts to be joined. As an added bonus, the flux actually reduces the surface tension and makes the solder flow and spread out more easily. So, you should be using an SN60 or SN63 rosin core solder.

The Soldering Iron

A good soldering iron will heat the surfaces to be joined to about 500° Fahrenheit so the solder can melt and flow properly. To do this, the soldering iron's tip must be hot enough and have enough capacity to initially heat and keep the surfaces at the proper temperature. To do this effectively, you'll need a soldering iron with at least a 700° Fahrenheit soldering element and a tip with enough mass to provide the heat needed for the job. The tip should be kept clean and free of debris and oxidation by initially

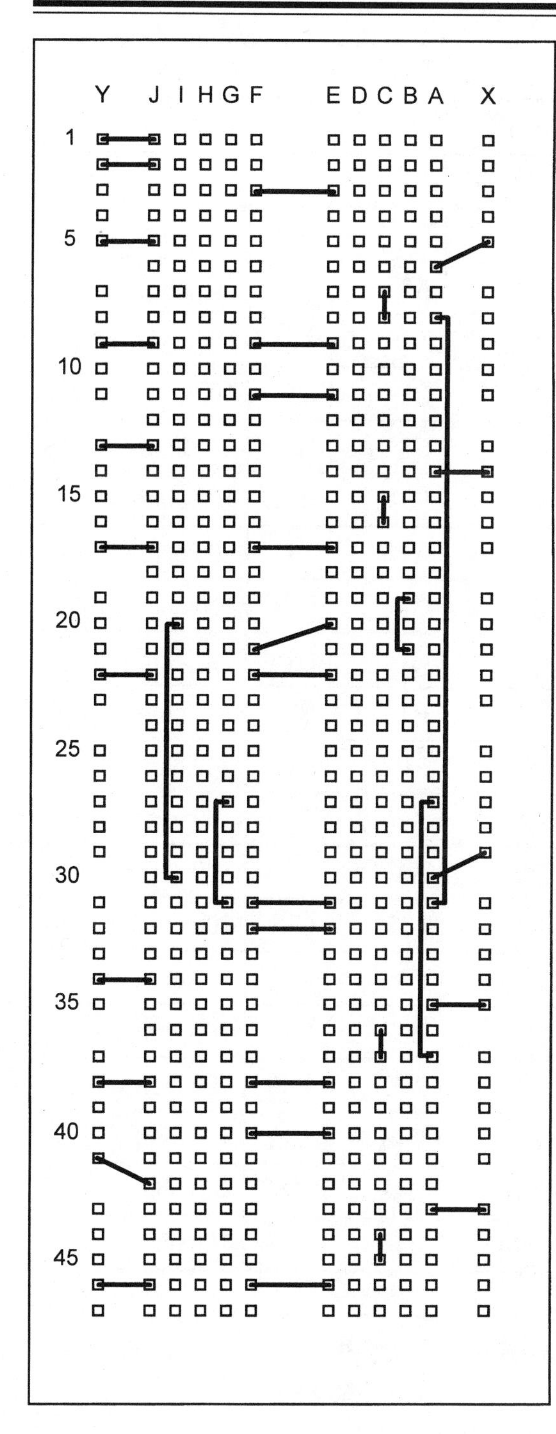

NAME	FROM	TO	REMARKS
JU1	Y1	J1	
JU2	Y2	J2	
JU3	F3	E3	
JU4	Y5	J5	
JU5	A6	X5	
JU6	C7	C8	
JU7	Y9	J9	
JU8	F9	E9	
JU9	F11	E11	
JU10	Y13	J13	
JU11	A14	X14	
JU12	C15	C16	
JU13	Y17	J17	
JU14	F17	E17	
JU15	B19	B21	
JU16	I20	I30	Insulated
JU17	F21	E20	
JU18	Y22	J22	
JU19	F22	E22	
JU20	G27	G31	
JU21	A27	A37	Insulated
JU22	A30	X29	Under JU21
JU23	F31	E31	
JU24	F32	E32	
JU25	Y34	J34	
JU26	A35	X35	Under JU21
JU27	C36	C37	
JU28	Y38	J38	
JU29	F38	E38	
JU30	F40	E40	
JU31	Y41	J42	
JU32	A43	X43	
JU33	C44	C45	
JU34	Y46	J46	
JU35	F46	E46	
JU36	A8	A31	Insulated

***Figure and Table 5-1**. Jumper placements.*

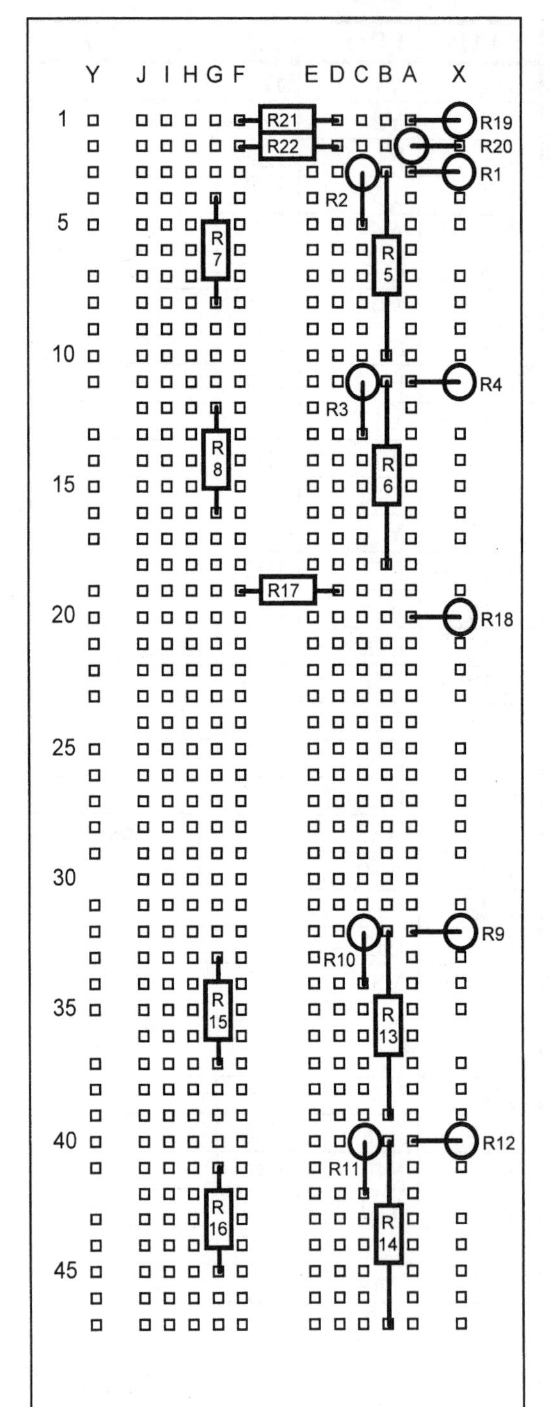

NAME	FROM	TO	REMARKS
R21	F1	D1	2.2kohm
R19	A1	X1	2.2kohm vertical
R22	F2	D2	2.2kohm
R20	A2	X2	2.2kohm vertical
R1	A3	X3	100 ohm vertical
R2	C3	C5	100 ohm vertical
R7	G4	G8	2.2kohm
R5	B3	B10	100 ohm
R4	A11	X11	100 ohm vertical
R3	C11	C13	100 ohm vertical
R8	G12	G16	2.2kohm
R6	B11	B18	100 ohm
R17	F19	D19	2.2kohm
R18	A20	X20	100 ohm vertical
R9	A32	X32	100 ohm vertical
R10	C32	C34	100 ohm vertical
R15	G33	G37	2.2kohm
R13	B32	B39	100 ohm
R12	A40	X40	100 ohm vertical
R11	C40	C42	100 ohm vertical
R16	G41	G45	2.2kohm
R14	B40	B47	100 ohm

***Figure and Table 5-2**. Resistor placements.*

tinning it (coating it with solder) and wiping it on a damp sulfur-free sponge just prior to applying it to the solder joint.

Solderability

As soon as you expose metal to air, it begins creating an oxide surface layer. As this oxide increases, the ability to solder properly (solderability) begins to decrease. So, it's a good idea to clean the copper side of the PCB with a light abrasive pad just prior to soldering.

When you create a joint that's not quite perfect, and continue to heat it to make it better, you may actually be making it worse! The problem is that you are burning off the flux which helps make a good solder joint. If you need to reheat a joint more than once, it's good practice to add some additional flux (or, if you don't have liquid flux available, some more solder with a rosin flux core). This instantaneously increases the joint's solderability and will always provide a better result.

Making the Joint

Presuming your soldering iron has a tip that is small enough to make good contact with all surfaces, but large enough to maintain a constant temperature, you're ready to begin. Referring to *Figure 5-6*, place the soldering iron tip in contact with both the component lead and the circuit board copper, positioning the tip so as much of its area contacts those two surfaces. Hold the tip firmly against the surfaces for about two seconds, and then feed the solder in from the side (not directly onto the tip). Once the solder begins to spread across the circuit board copper, continue to hold the tip in place for about another second. Then remove the tip and leave the joint undisturbed for at least two seconds. Once the joint has cooled, the excess component lead can be clipped flush with the top of the solder joint. Since the component lead is mechanically secured above the solder joint, clipping now avoids placing undue stress on the component/component lead interface (the mechanical vibration from the cutting process is absorbed by the joint, not transferred into the component).

Wire Preparation

When properly prepared and soldered, wire makes a good electrical and mechanical joint. Preparation begins by cutting it to the desired length and then stripping an appropriate amount of insulation from the wire ends. The standard stripping length for wires

NAME	FROM	TO	REMARKS
Q5	H3	H5	Note 1
Q1	E5	E7	Note 2
Q3	D8	D10	Note 3
Q6	H11	H13	Note 1
Q2	E13	E15	Note 2
Q4	D16	D18	Note 3
Q14	H20	H22	Note 1
Q13	D20	D22	Note 1
K1	F23	E30	Note 4
Q11	H32	H34	Note 1
Q7	E34	E36	Note 2
Q9	D37	D39	Note 3
Q12	H40	H42	Note 1
Q8	E42	E44	Note 2
Q10	D45	D47	Note 3

NOTE 1: PN2222. Flat towards X.

NOTE 2: **MJE**3055. Tab towards X. Bend towards Y to clear adjacent transistor.

NOTE 3: MPS2955. Tab towards Y.

NOTE 4: Mount with Pin 1 at E30.

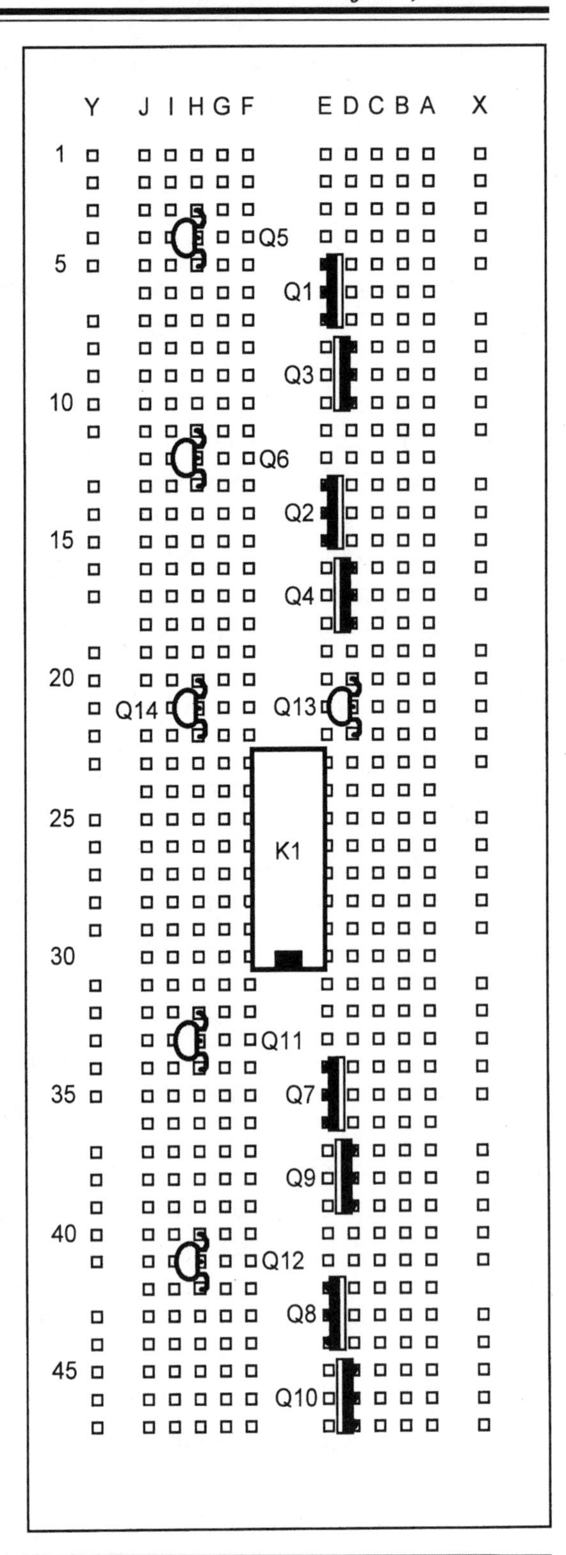

***Figure and Table 5-3**. Transistor and relay placements.*

Name	From	To	Remarks
S2G	H1	S2	GND of S2
S2+	C1	S2	N/O of S2
S1G	H2	S1	GND of S1
S1+	C2	S1	N/O of S1
P11	B1	P1-11	P1-11
P10	B2	P1-10	P1-10
GND	Y4	GND	From Power Suppply
V+	X4	+5	From Power Supply
P5	J8	P1-5	P1-5
P19	Y14	P1-19	P1-19 (Gnd)
P4	J16	P1-4	P1-4
P9	J19	P1-9	P1-9
P3	J37	P1-3	P1-3
P2	J45	P1-2	P1-2
M2A	J23	M2-A	Drill
M1A	J25	M1-A	Table
M2B	A23	M2-B	Drill
M1B	A25	M1-B	Table
M2B*	B16	M2-B*	Drill
M1B*	A16	M1-B*	Table
M2A*	B45	M2-A*	Drill
M1A*	A45	M1-A*	Table

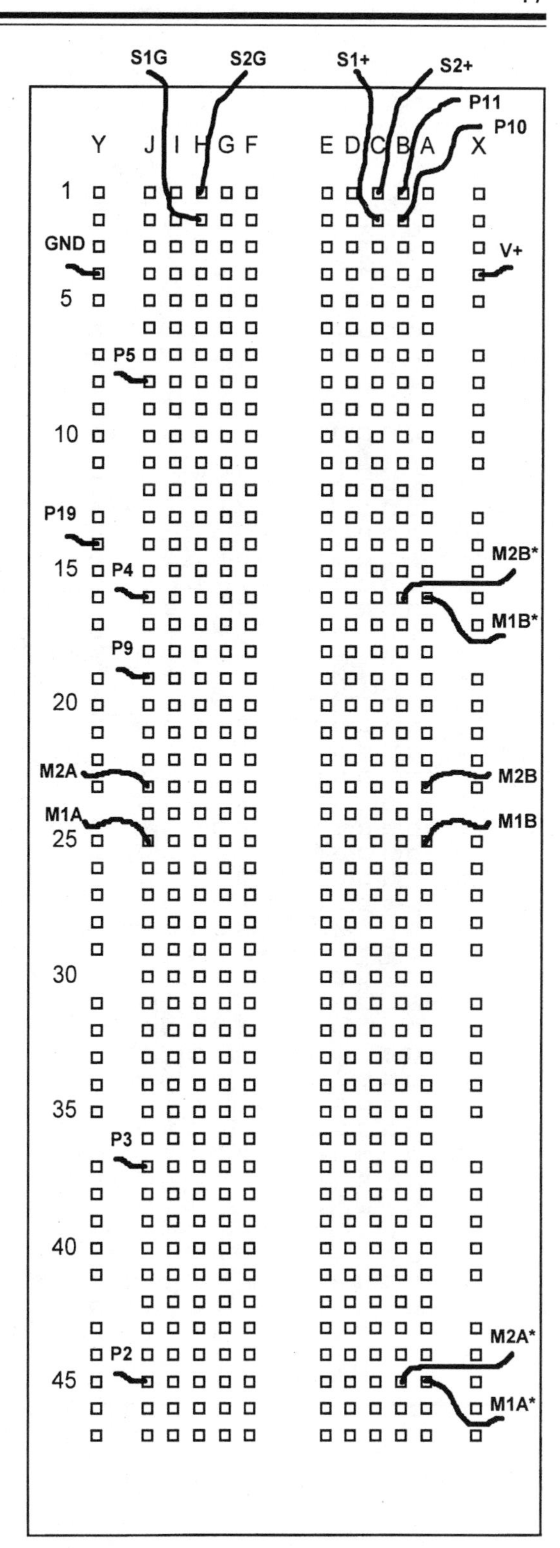

Figure and Table 5-4.
Wire connections.

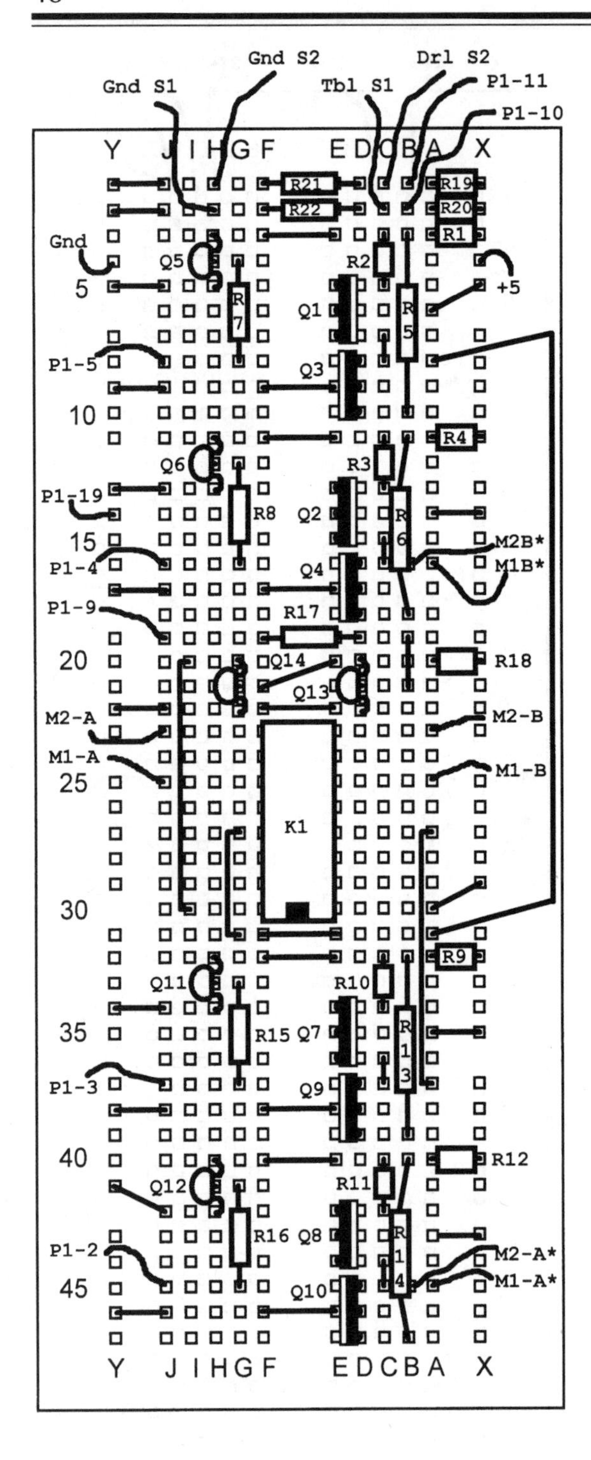

***Figure 5-5**. The composite board.*

to be attached to a circuit board is 1/4". This allows at least 1/8" of bare wire between the area to be soldered and the insulation. This "safe zone" helps avoid melting the insulation during soldering, and solder traveling up the wire strands by capillary action. Called solder "wicking," this action not only brings solder up the strands, but also flux and contaminants which can be trapped under the insulation. Over time, this debris can cause failure of the joint.

The most common mistake when stripping wire is to nick the metal strands while trying to remove the outer insulation. Cut or nicked wire strands weaken the wire and the resultant joint, and are most often created by using a diagonal pliers to strip the wire instead of a wire stripping tool.

Stripping Tools

A wire stripper is preferable over any other cutting tool. It serves two purposes; it allows a precise amount of insulation to be removed, and it minimizes damage to the solid wire or wire strands. A good wire stripper will clamp down on the wire, piercing the insulation but not the wire inside. The best wire stripper will have graduated cutting areas for specific sizes of wire.

Tinning

Once stranded wire has been stripped, it should be tinned. Tinning deposits solder onto and in between the strands, keeping the individual strands from separating and avoiding potential shorts when the wire is attached to its destination.

To tin, twist the strands clockwise forming a tight bundle about the same diameter as the untwisted wire. Place the soldering iron tip farthest from the insulation, and feed solder towards the tip. As the solder begins to melt, move the solder towards the insulation, stopping about 1/8" from the insulation. When all strands are coated, remove both the solder and the iron.

Uninsulated Jumpers

A jumper is a wire connection from one electrical point to another. An uninsulated jumper is made from a piece of bare wire (like the excess lead that is clipped off after installing a resistor or capacitor on a circuit board, or a solid hookup wire that has had its insulation stripped off). To form a jumper, place it in the jaws of a needle nosed pliers and, while holding the jaws closed, bend both ends down to form a "U." This project will use #22 solid wire to form uninsulated jumpers.

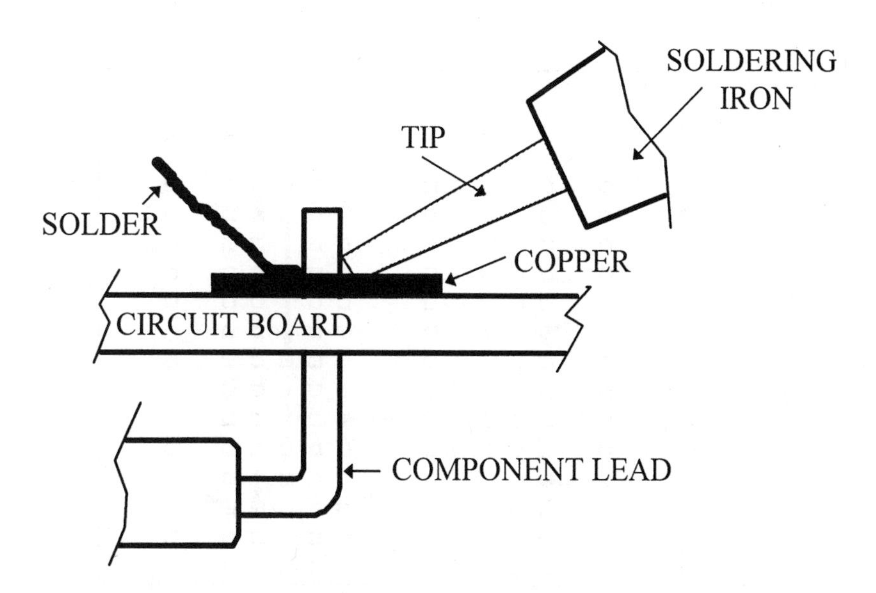

***Figure 5-6**. Place the soldering iron tip in contact with both the component lead and the circuit board copper.*

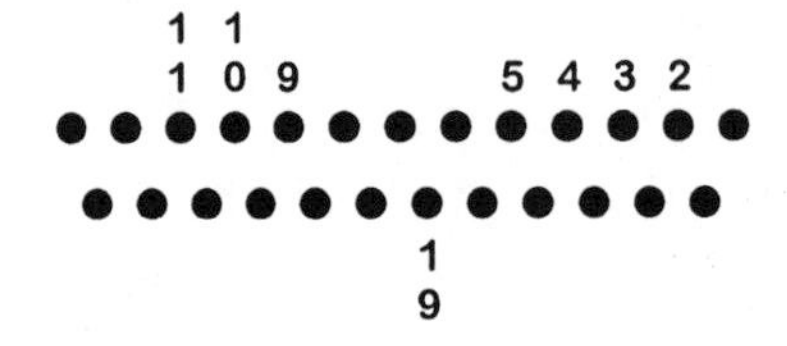

Figure 5-7. Pinouts on the DB-25 connector.

Insulated Jumpers

An insulated jumper is a length of solid or stranded hookup wire whose ends have been stripped of insulation. Unlike wires attached to circuit boards, the strip length should be 1/8" on each end. This project will use #22 solid wire to form the three required insulated jumpers identified as "Insulated" in the Remarks column of *Table 5-1*.

Forming Vertical Mount Resistors

Resistors R1-R4, and R9-R12 are indicated in the Remarks column of *Table 5-2* as "Vertical". The mount spacing for these resistors is shorter than the resistor body, so mount the resistor upright in one of the mounting holes and solder in place. Then, bend the free resistor lead down and into the other mounting hole.

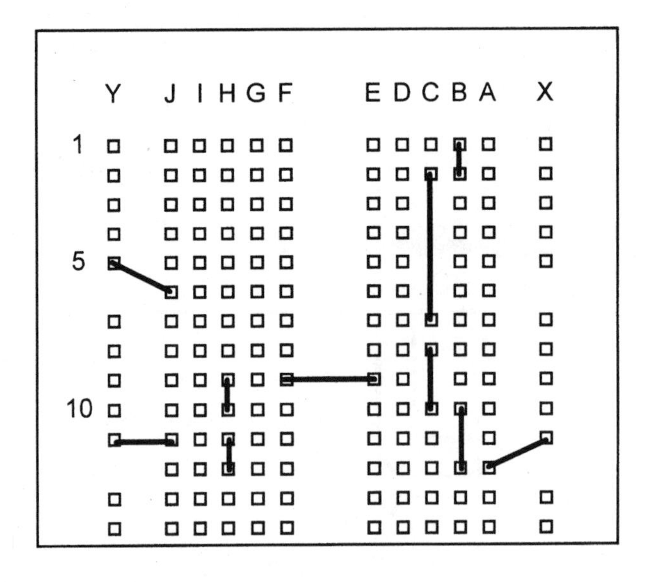

Figure 5-8. Uninsulated jumper placements.

NAME	FROM	TO
JU1-P	B1	B2
JU2-P	C2	C7
JU3-P	Y5	J6
JU4-P	H9	H10
JU5-P	F9	E9
JU6-P	C8	C10
JU7-P	Y11	J11
JU8-P	H11	H12
JU9-P	B10	B12
JU10-P	A12	X11

Table 5-5. *Uninsulated jumpers for the power supply PCB.*

The PCBs

The solderless breadboard type PCB specified in *Table 3-1* duplicates the hole layout of a standard solderless breadboard. It has a copper and non-copper (component) side. On the PCB's component side are twelve columns labeled (from left to right) Y, J, I, H, G, F, E, D, C, B, A, and X. The PCB also has 47 rows, with rows 1, 5, 10, 15, 20, 25, 30, 35, 40 and 45 labeled.

The controller board will use one of these PCBs. The power supply PCB will use a segment of another PCB. That segment will be obtained by cutting the required portion from a full-sized PCB.

The DB-25 Connector

Table 5-4 and *Figure 5-4* show connection between the PCB and off-board components. A number of wires are referenced as Px (like P2 or P19). These refer to the pin numbers on a DB-25 male (plug) connector whose pin identification (looking at the rear, or solder cups, of such a DB-25 connector) is shown in *Figure 5-7*. Individual lengths of #22 or #24 stranded wire may be used to connect between the PCB and the DB-25 connector. Alternately, a suitable length (up to ten feet is okay) of a 9 (or more) conductor cable can be used to connect between the PCB and the DB-25 connector.

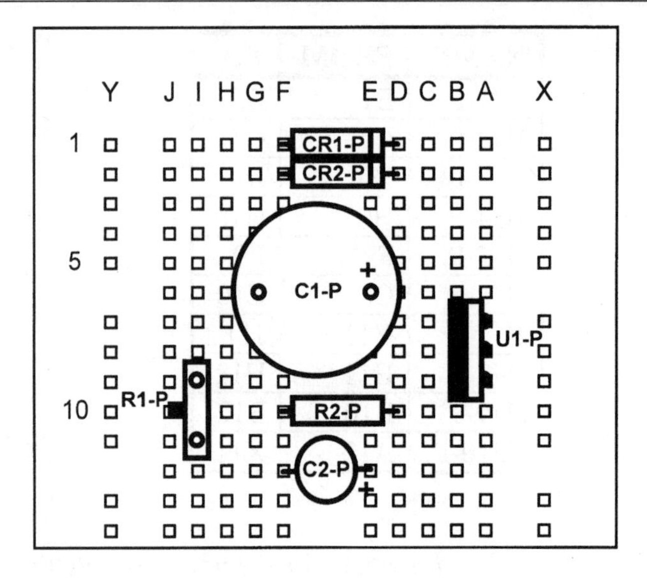

***Figure 5-9**. Component placements.*

Installing the Jumpers on the Controller Board

Referring to *Table 5-1* and *Figure 5-1*, form and install the 33 uninsulated and 3 insulated jumpers. After completion, go back and check the position of each end of each jumper.

Installing the Resistors on the Controller Board

Referring to *Table 5-2* and *Figure 5-2*, install the 22 resistors. Note that 11 of those resistors are installed vertically. Before installing each resistor, check the color code (brown-black-brown for 100 ohms, and red-red-red for 2.2 kohms) to ensure you are installing the correct value resistor.

Installing the Active Components on the Controller Board

Referring to *Table 5-3* and *Figure 5-3*, install the six PN2222 transistors, four MJE2955T transistors, four MJE3055T transistors, and one 16 pin DIP relay. Note that the PN2222 transistor has a cylindrical body with a flat. The *Table 5-3* notes refer to the flat to ensure proper orientation of the transistor. Also note that the MJE2955T and MJE3055T transistors have a rectangular body with a larger metal heat sink tab. The *Table 5-3* notes also refer to the tab to ensure proper orientation of those transistors. As indicated in Chapter 2, these transistors will not fit side-by-side without some adjustment. Install Q1 and then push it toward the center of the PC board, bending the leads in the process. This provides sufficient clearance to allow Q3 to fit. Repeat this for Q2, Q7, and Q8. Finally, note that the relay K1 is installed with pin 1 nearest the orientation marking in *Figure 5-3*.

Controller Board Final Wiring

Referring to *Table 5-4* and *Figure 5-4*, you will now make 20 wire connections from the PCB. Eight of these connections will be to a DB-25 connector previously discussed. The other 12 connections consist of 4 for each of the two stepper motors, and 2 for each of the limit switches. (The two connections shown for GND and +5V will be made later, with wires coming from the power supply PCB.) Prepare 12 two-foot lengths of #22 or #24 stranded wire, and an appropriate length of 9 conductor cable. Connect the wires and 8 of the 9 conductors in the cable to the identified points on the PCB. Connect the 8 conductors in the cable to the appropriate points on the DB-25 connector, referring to *Figure 5-7* for the connector pinouts. Leave all other 12 wires unconnected at this time. *Figure 5-5* provides a view of the completed PCB with all jumpers, components, and wires installed.

NAME	FROM	TO	REMARKS
CR1-P	F1	D1	Band towards row x
CR2-P	F2	D2	Band towards row x
C1-P	G6	E6	+ at E6
C2-P	F12	E12	+ at E12
R1-P	I9	I11	Center at J10
R2-P	F10	D10	100 ohms
U1-P	A7	A9	Tab Facing C1-P

Table 5-6*. Components for the power supply PCB.*

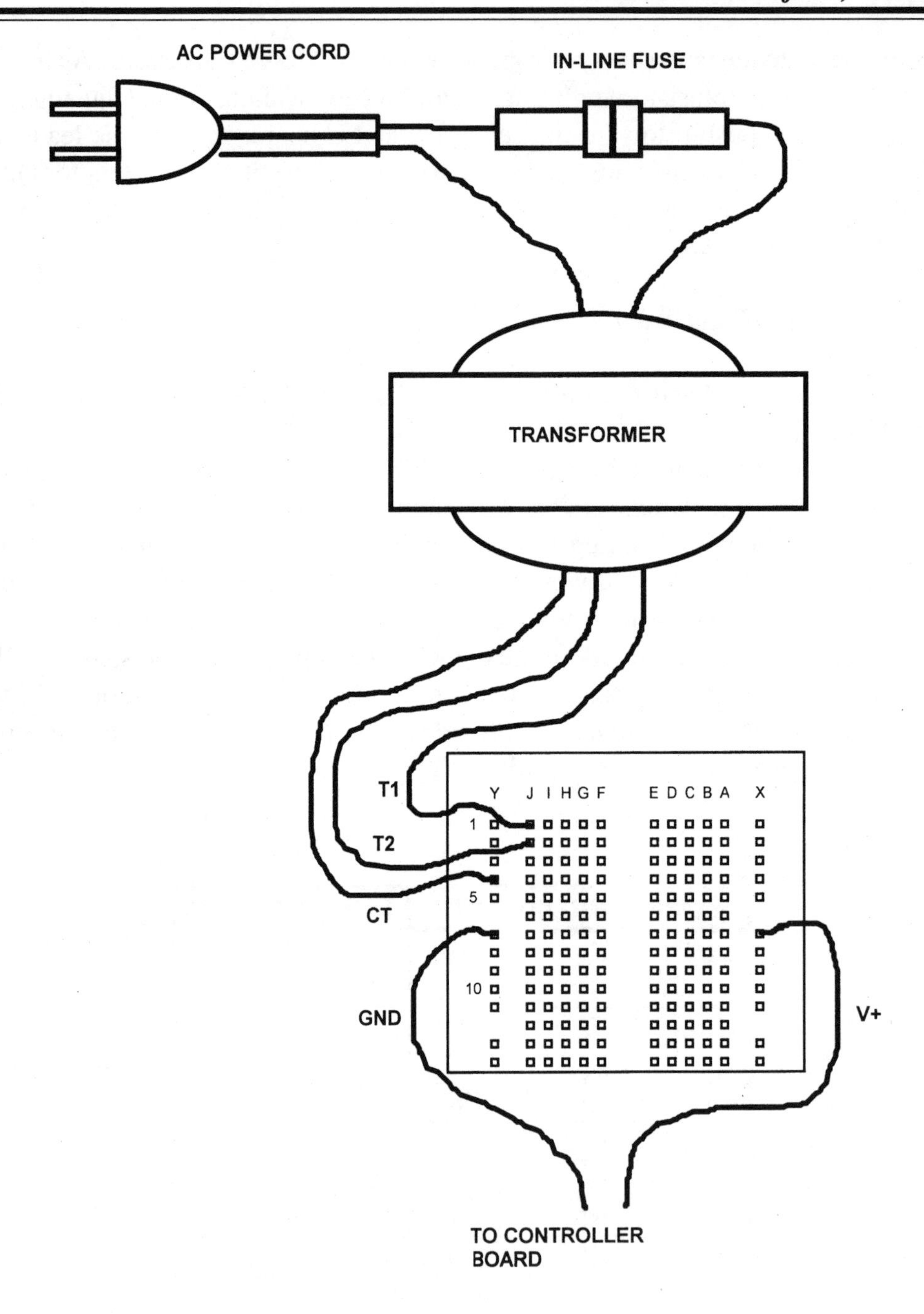

Figure 5-10*. Power supply wiring placements.*

The Power Supply Board

The power supply board is a segment of one solderless breadboard type PCB specified in *Table 3-1*. To create this PCB, locate row 15. Using a hacksaw or similar cutting tool, cut through this row to form a PCB with 14 rows. When done, file the cut edge smooth.

Installing Jumpers on the Power Supply PCB

Referring to *Table 5-5* and *Figure 5-8*, install the ten uninsulated jumpers on the PCB. When done, go back and check the position of each end of each jumper.

Installing the Components on the Power Supply PCB

Referring to *Table 5-6* and *Figure 5-9*, install the two diodes, two capacitors, PC mount trimmer potentiometer, fixed resistor, and LM317T voltage regulator IC onto the PCB. Note the position of the positive lead of both capacitors and the orientation band on CR1-P and CR2-P. Also note the orientation of the PC mount trimmer potentiometer. Note that the LM317T has a rectangular body with a larger metal heat sink tab. The metal tab should be closest to the center of the PCB. After installation, install the heat sink onto the LM317T by using the hardware provided with the heat sink, or with a #4-40 x 1/2" machine screw and #4-40 nut.

Power Supply Final Wiring

Referring to *Table 5-7* and *Figure 5-10*, connect the three secondary leads from the transformer to the appropriate points on the PCB. Install a 6" length of wire into the GND and V+ positions on the PCB. Connect one primary transformer lead to one lead

NAME	FROM	TO	REMARKS
T1	J1	T1	Transformer
T2	J2	T2	Transformer
CT	Y4	CT	Transformer
GND	Y7		Controller Board
V+	A7		Controller Board

Table 5-7. *Power supply wiring.*

from the AC power cord. Connect the other primary transformer lead to one of the leads of the in-line fuse holder. Connect the other in-line fuse holder lead to the remaining lead of the AC power cord. Protect all AC connections with either wire nuts or shrink sleeving to avoid any possibility of shock. Install a 250V, 1A fuse in the in-line fuse holder.

Testing

You'll now test the power supply to ensure it is working correctly, and then adjust the voltage:

1. Connect the V+ lead from the power supply PCB to the positive lead of your voltmeter.
2. Connect the GND lead from the power supply PCB to the negative lead of your voltmeter.
3. Set the voltmeter to the 10 volt range. (If your voltmeter doesn't have a 10 volt range, set it to a range that will allow you to measure at least 7 volts.)
4. Set potentiometer R1-P to its mid range.
5. Apply AC power.
6. You should measure a voltage somewhere between 3 and 13 volts DC. If not, unplug the power supply and check for incorrectly installed components, mis-wirings, etc. Correct any problems and try again.
7. Adjust R1-P for a voltage of 5.7 to 6.0 volts.

When you have successfully tested the power supply, connect the V+ lead from the power supply PCB to position X4 on the controller PCB. Connect the GND from the power supply PCB to position Y4 on the controller PCB. As a reference, *Figure 5-11* shows the completed power supply PCB with all jumpers, components, and wires installed.

Now that you've successfully completed construction of the control and power supply electronics, it's time to test them out. Here's what you'll need:

1. Control and power supply electronics.
2. Four jumper leads with alligator clips on both ends.
3. Your PC.
4. This book's companion software disk.
5. A stepper motor.

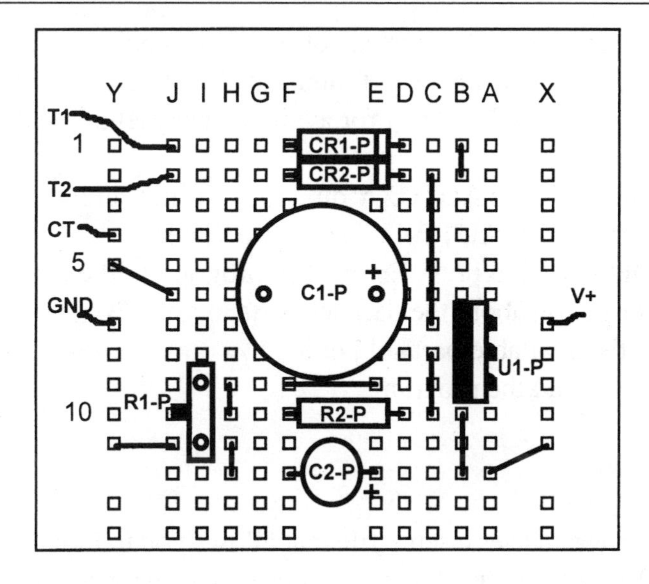

Figure 5-11. *The completed power supply PCB with elements installed.*

Loading the Software

You'll now load the software from the floppy disk onto your PC. To do that, perform the following steps:

1. Start up your PC and enter the DOS mode. (For Windows '95 users, select "Start," then "Programs," then "MS-DOS Prompt.")
2. Assuming your hard drive is C:, type "CD C:\" then press Enter to move to the hard drive's root directory.
3. Type "MD AUTOXY" and press Enter to create a new directory called AUTOXY.
4. Type "CD AUTOXY" and press Enter to move to that new directory.
5. Install the floppy disk provided into your floppy drive.
6. Assuming your floppy disk is A:, type "COPY A:*.*" and press Enter to copy all files from the floppy disk to the new AUTOXY directory.

Checking for Available Parallel Ports

The floppy disk contains a program called "FINDPORT.EXE." This program, which is now on your AUTOXY, directory checks for the presence of parallel ports 1 through 4, verifies that they are responding, and gives you the decimal and hexadecimal addresses of each port found. If a port is not installed, you'll receive the message "Not Found" for that port. If a port was found but is not responding, you'll receive the message "Not Responding." To test for available parallel ports, do the following:

1. Make sure you are in the AUTOXY directory. (If not sure, type "CD C:\" and press Enter. Then type "CD AUTOXY" and press Enter.)
2. Type "FINDPORT" and press Enter. The program will execute and provide you with the information about the parallel ports in your PC. Write down the decimal addresses of the available parallel ports.
3. Press any key to end the program.

Making a Setup File

The AUTOXY program uses a setup file that allows you to modify various parameters like the parallel port to be used, the speed at which the stepper motors will run, the factor related to the pitch of the threaded rods you're using, and whether a startup message should appear. To create the setup file do the following:

1. Make sure you are in the AUTOXY directory. (If not sure, type "CD C:\" and Press Enter. Then type "CD AUTOXY" and press Enter.)
2. Type "COPY CON DRLSETUP.DAT" and press Enter.
3. Assuming the parallel port you want to use is at decimal address 888, type "888" and press Enter. If the parallel port you want to use is at another address (like 928 decimal), type its number instead of 888.
4. Type "1000" and press Enter.
5. Type "864" and press Enter.
6. Type "Message" and press Enter.
7. Hold down the Ctrl key, press the Z key, then release both keys. The screen will show "^Z."
8. Press Enter. You will see the message, "1 file(s) copied."

Starting the SPEEDSET Program

Now that you've created the required setup file, you can start the SPEEDSET program. To start up the program, do the following:

1. Make sure you are in the AUTOXY directory. (If not sure, type "CD C:\" and press Enter. Then type "CD AUTOXY" and press Enter.)

```
Current Speed Factor:  1000

  PRESS      TO
  -----      -----------------------------
    2        Move Drill Stepper Clockwise
    4        Move Table Stepper Clockwise
    6        Move Table Stepper Counterclockwise
    8        Move Drill Stepper Counterclockwise
   Esc       End
  Enter      Change Speed Factor
```

***Figure 5-12**. SPEEDSET program screen.*

2. At the DOS prompt, type "SPEEDSET" and press Enter. You will see the SPEEDSET program screen, as shown in *Figure 5-12*.

Take a look at this screen. You can move either the table (M1) or drill (M2) stepper motors in either direction with the four "motion" keys (2, 4, 6 or 8). When you press any of the motion keys, a blinking message "Press Any Key to Stop" will appear at the bottom of the screen. Press any key when you want to stop motion.

Pressing the Esc key brings up another message asking you if you want to save the new speed data. Press Y if you do, or N if you don't. If you press Y the current speed data will be saved in the DRLSETUP.DAT file. In either event, the program will end.

The speed factor is a timing delay between subsequent commands to the stepper motor. The lower the factor, the faster the motor will run. If the speed factor is set too low, the motor will not be able to respond and will simply "chatter" as it tries to move, but can't. If the speed factor is too high, it will take a very long time between steps. A suggested starting point for speed factors is:

1. 4.77 MHz PC: 1
2. 8-12 MHz PC: 10
3. 20-33 MHz PC: 100
4. 66-100 MHz PC: 1000
5. Above 100 MHz PC: 10000

With the motors free running (and not attached to the table or drill caddy), they will be able to run using a lower speed factor. When you attach them, friction and other factors will require that you run this program again to determine the optimum factor.

Testing the Table Motor Control Elements

You'll now temporarily connect a stepper motor to the table motor (M1) control electronics and verify proper operation of the motor and its limit switch (S1) circuitry. To perform this verification, do the following:

1. Identify the motor's A, A*, B, and B* leads.
2. Connect the M1-A lead from the controller PCB (position J25) to the motor's A lead.
3. Connect the M1-B lead from the controller PCB (position B25) to the motor's B lead.
4. Connect the M1-A* lead from the controller PCB (position A45) to the motor's A* lead.
5. Connect the M1-B* lead from the controller PCB (position A16) to the motor's B* lead.
6. Make sure the two leads for switch S1 (positions C2 and H2) are not touching. Position the stripped, free ends of these leads in a convenient place.
7. Apply power to the power supply.
8. Press the 4 key on the keypad (or on the main keyboard). Note the stepper motor moves in the clockwise direction as you view the shaft head on. Short the two S1 leads to each other and note that the motor continues to move. Press any key to stop movement.
9. If the motor is moving very slowly, decrease the speed factor until the motor no longer moves and simply "chatters". Then increase the speed factor a little until the motor move freely again.
10. Press the 6 key on the keypad (or on the main keyboard). Note the stepper motor moves in the counterclockwise direction as you view the shaft head on. Short the two S1 leads to each other and note that the motor stops immediately.
11. Connect the M2-A lead from the controller PCB (position J23) to the motor's A lead.
12. Connect the M2-B lead from the controller PCB (position B23) to the motor's B lead.
13. Connect the M2-A* lead from the controller PCB (position B45) to the motor's A* lead.

14. Connect the M2-B* lead from the controller PCB (position B16) to the motor's B* lead.
15. Make sure the two leads for switch S2 (positions C1 and H1) are not touching. Position the stripped, free ends of these leads in a convenient place.
16. Apply power to the power supply.
17. Press the 8 key on the keypad (or on the main keyboard). Note the stepper motor moves in the counterclockwise direction as you view the shaft head on. Short the two S2 leads to each other and note that the motor continues to move.
18. Press the 2 key on the keypad (or on the main keyboard). Note the stepper motor moves in the clockwise direction as you view the shaft head on. Short the two S2 leads to each other and note that the motor stops immediately.

Of course, if you do not get these responses, go back and recheck your work. Correct any mistakes and try again.

Wrapping it Up

Now that we've completed the control and power supply electronics and successfully tested them, it's time to move on to constructing the other parts of the *Auto-XY* project. In Chapter 6, we'll construct the table, its supports, and the mechanism that holds the table stepper motor in place.

CHAPTER 6
Building the Table Assembly

Introduction

Now that we've completed the *Auto-XY* case and electronics, it's time to move on to the first of the two major assemblies; the table. It consists of a flat wooden table surface on which is attached the table drive and the limit switch trip arm. The table surface rides on two parallel tracks, is kept moving straight by the table guide, and is driven by the table stepper motor. The table stepper motor is secured to the *Auto-XY* base with two table stepper mounts.

In this chapter, we'll fabricate the table surface, the table drive, the table guide, the limit switch trip arm, the table tracks, and the table stepper mounts. When we're finished assembling these items to the base, we'll adjust operation using the SPEEDSET program. Many of the processes we go through in this task will be similar to those we'll need for the next step; the drill assembly. As we proceed, you'll need to refer back to the tables in Chapter 3 for the dimensions and other details on the individual parts.

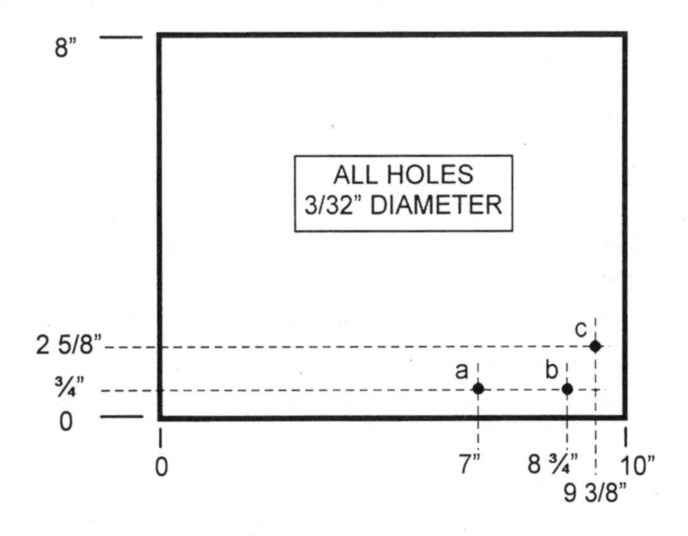

Figure 6-1*. Drill holes in the plywood as shown.*

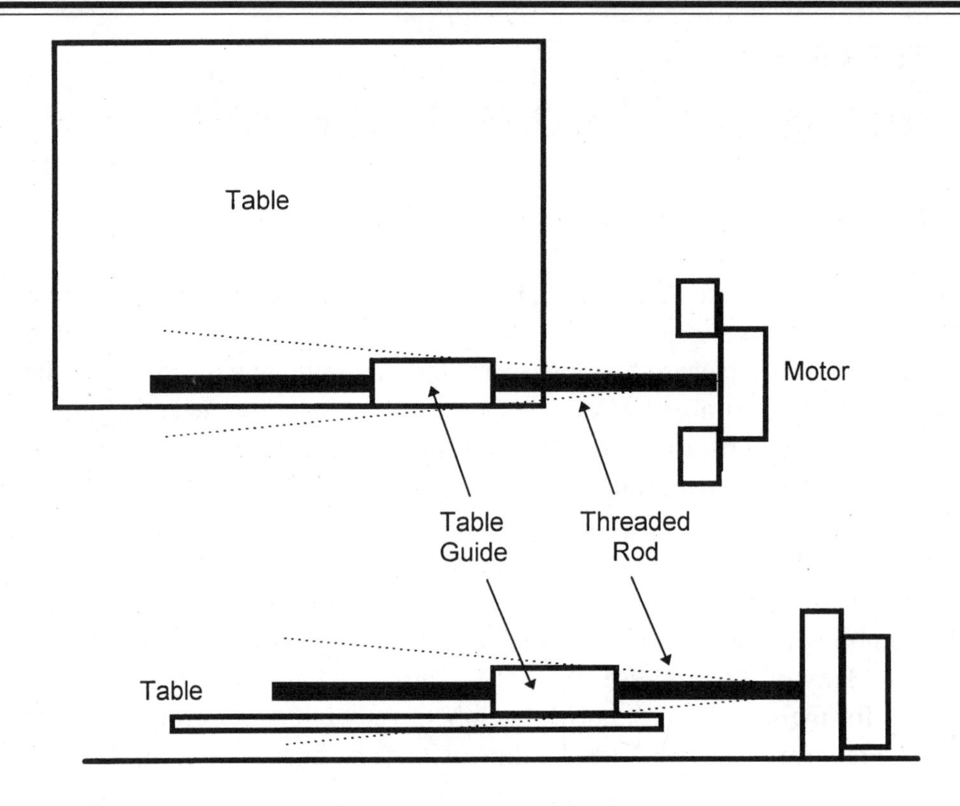

Figure 6-2. *The dotted lines show how the table should move parallel to the plane of the stepper motor's shaft.*

The Table Surface

The table surface (#41) is an 8" x 10" piece of 1/4" plywood. When forming this item, be sure to make the four sides square. Once you have cut the plywood to size, drill holes *a*, *b*, and *c* as shown in *Figure 6-1*. Holes *a* and *b* will allow the table driver to be mounted on the top of the table surface. Hole *c* allows the limit switch trip arm to be mounted on the bottom of the table surface. After drilling, sand all surfaces smooth and finish with water based varnish or other suitable clear coating. Let dry thoroughly before proceeding.

The Table Driver

The table driver is a critical table assembly component. It accepts the threaded rod and moves the table back and forth in the X direction as the table stepper motor rotates

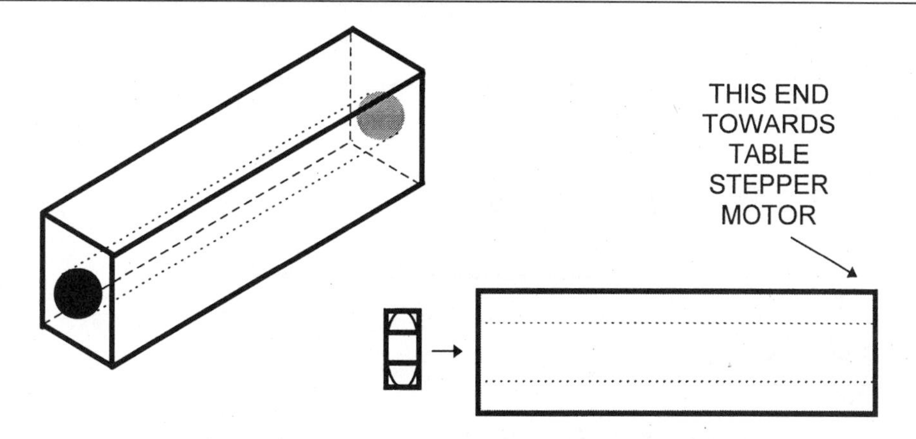

***Figure 6-3**. Various views of the driver showing where the nut is to be inserted.*

clockwise and counterclockwise. Referring to *Figure 6-2*, if the driver does not allow the table to move parallel to the plane of the stepper motor's shaft (as shown in an exaggerated fashion by the dotted lines), either the table will wobble, or the increased torque required will tend to stall the motor. This is true for side-to-side and up-and-down directions.

To see which portions of the table driver are most critical, let's look at the actual component. *Figure 6-3* shows a three-dimensional view of the driver (#43) prior to installation of the drive nut (#18), a side view of the driver, and how the nut is inserted. In a perfect world, you would drill an appropriately sized hole through the driver that is exactly parallel to all of its sides. Then you would press a nut in one end, the end being exactly perpendicular to all of the sides. In this way, you would have a driver that, when mounted to the table surface, would avoid any misalignment.

Fabricating the Table Driver

In our not-so-perfect world, the hole will not be quite centered and slightly out of parallel. However, this will not be a problem, since the position of the nut in the driver can be adjusted slightly to correct for any misalignment. To fabricate the table driver, do the following:

1. Cut a 1" wide x $2^{1/4}$" long piece of wood from 3/4" stock (#43).
2. Locate the center of each short end by drawing diagonals from opposing corners. Using a center punch, mark each center.

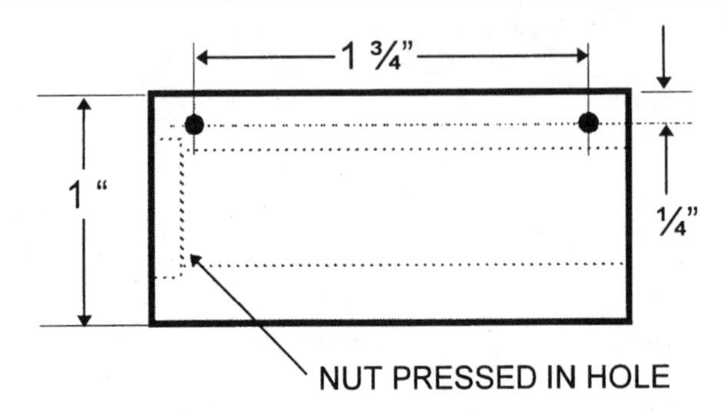

Figure 6-4. Drill the two 3/32" holes as shown.

3. Using a drill press with a 3/32" bit and some form of guide, hold the driver blank on its end and parallel to the drill bit. Drill the hole about half way through the blank. Turn the blank end-for-end and drill the other center-punched hole.
4. Using successively larger bits, continue this process until you have a through-hole that is slightly smaller than the outside dimensions of the drive nut you'll be using. (For the specified 5/16" threaded rod, that final drill bit size should be 1/2".)
5. Place a nut on the end of the blank, approximately centered in the hole. Temporarily hold it there with masking tape.
6. Insert the assembly into a vise and *slowly* press the nut into the hole.

Final Drilling and Assembly

The last step in fabricating the table driver is to drill two 3/32" holes as shown in *Figure 6-4*. These two holes will mate with the two 3/32" holes drilled in the table (*Figure 6-1*). Sand all surfaces smooth and finish with water based varnish or other suitable clear coating. Let dry thoroughly before proceeding. When dry, thread two #4-40 x 3/4" screws (#3) through the bottom of the table surface and into the table driver. Make sure the screws are tightened enough to "snug up" the driver to the top of the table surface.

The Table Guide

The table guide is a short length of angled aluminum with an elongated hole drilled in one side. The guide fits on top of the table driver and straddles the table track. When adjusted, the guide prevents the table from moving sideways at is driven back and forth by the table stepper motor. To fabricate the table guide, do the following:

1. Cut a $2^{1/4}$" length of angle aluminum (#34).
2. Drill two 1/8" diameter holes in one surface of the table guide using the dimensions shown in *Figure 6-5a.*
3. Remove the material between the two holes with a file. The resulting guide should appear as depicted in *Figure 6-5b.*

Fabricate a second guide and retain in for later use on the drill assembly.

The Limit Switch Trip Arm

The limit switch trip arm attaches to the end of the table and contacts the plunger of the table limit switch S1 at the end of the table's movement toward the table stepper. Adjusting the position of this trip arm sets (or resets) the zero position of the table. To make and install the trip arm, do the following:

1. From a miscellaneous piece of scrap dimensional lumber (#60), cut a 2" long x 3/4" wide x 1/4" thick piece to form the trip arm.
2. Drill an elongated hole in this piece as shown in *Figure 6-6a.*
3. Sand all surfaces smooth and finish with water based varnish or other suitable clear coating. Let dry thoroughly before proceeding.

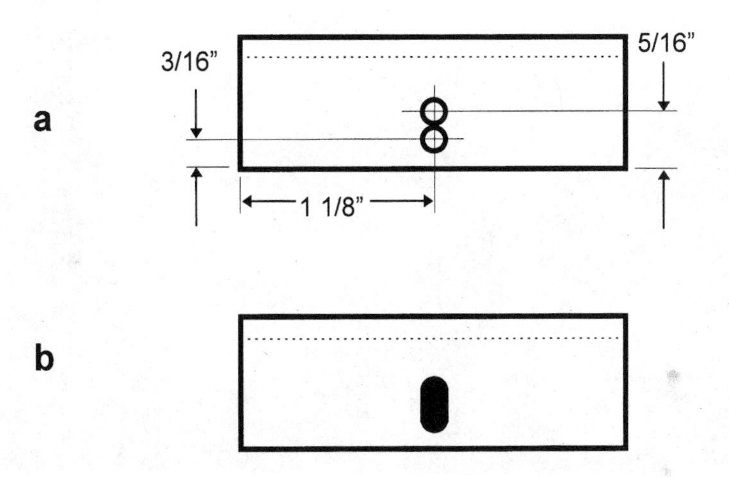

Figure 6-5. *Hole placements for the table guide.*

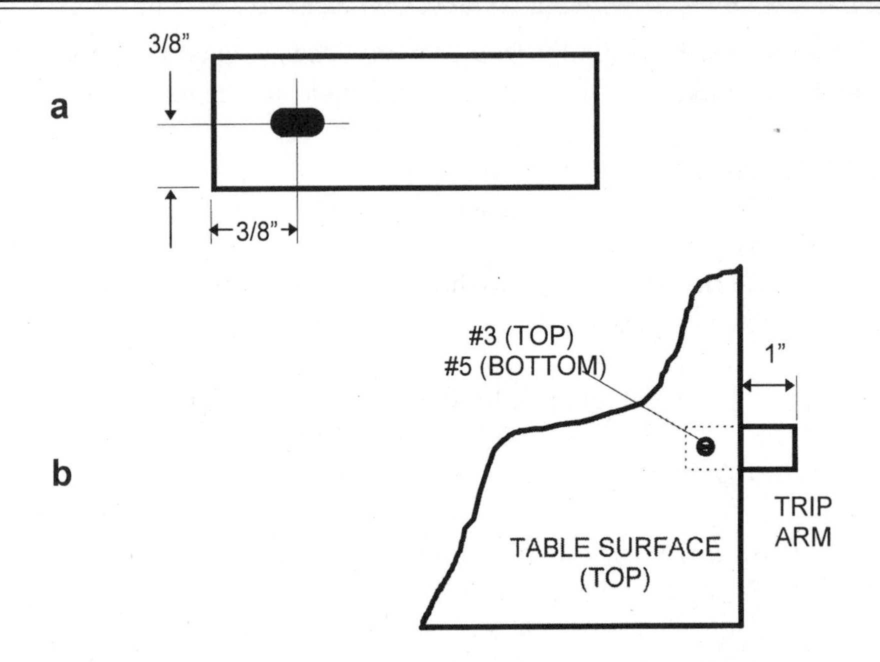

***Figure 6-6**. Hole placement and mounting of the trip arm.*

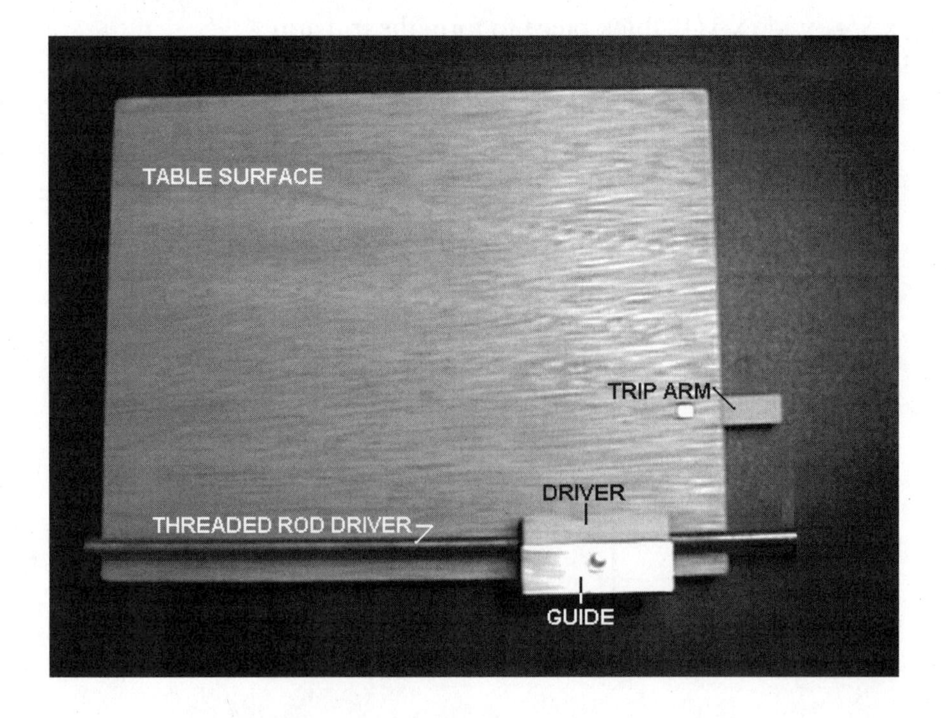

***Photo 6-1**. Completed table surface.*

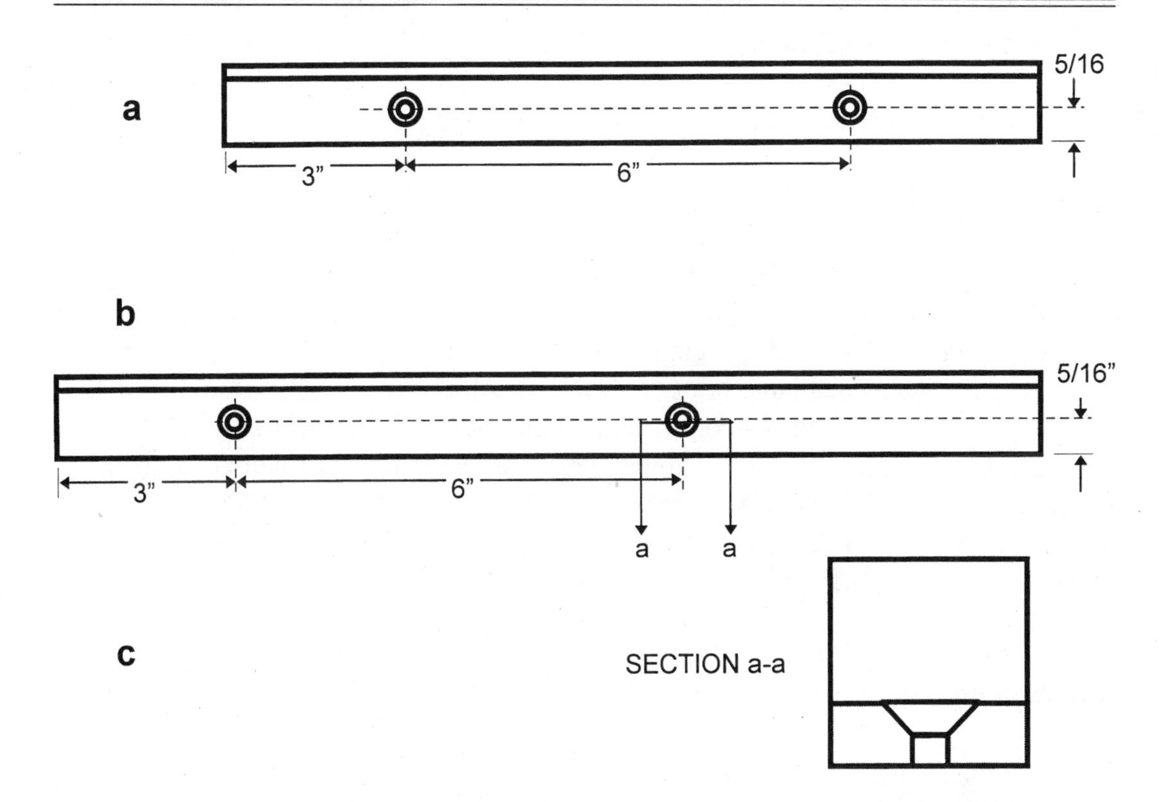

Figure 6-7. *Hole placements for the table tracks.*

When dry, mount the trip arm to the bottom of the table surface as shown in *Figure 6-6b* using a #4-40 x 3/4" machine screw (#3) inserted from the top of the table surface through the elongated hole in the trip arm. Secure the screw with a #4-40 nut (#5) so the trip arm protrudes about 1" from the edge of the table surface. (A finer adjustment will be performed during initial table alignment.) *Photo 6-1* shows the completed table surface with the table driver, table guide and limit switch trip arm installed (along with the threaded rod drive).

The Table Tracks

The table tracks are two lengths of aluminum angle stock that are secured to the base with machine screws. The table rides in these two tracks.

Fabricating the Table Tracks

To fabricate the table tracks, do the following:

1. Cut one 14" length of angle aluminum. This is the long table track (#39).
2. Cut one 12" length of angle aluminum. This is the short table track (#40).
3. Obtain four #8-32 x 1 1/4" flat head machine screws (#12). Drill two 5/32" diameter holes in the short track (*Figure 6-7a*) and two 5/32" diameter holes in the long track (*Figure 6-7b*) to accept the screws.
4. Measure the outside diameter of the head. (It should be about 5/16".) Use that diameter drill bit to form a countersink in the four holes just drilled. Countersink each hole to a depth that will allow the head to be flush or just below the track's surface. A cross section of the resulting hole is shown in *Figure 6-7c*.
5. Mount one screw in each of the four holes, securing each screw to the track with a #8 flat washer (#13) and an #8-32 nut (#14). Screw an additional nut (#14) onto each of the four screws.

Installing the Tracks on the Base

The machine screws previously installed in the table guides will be inserted through another #8 flat washer and into matching holes in the base. Then the loose nuts previously installed on the machine screws will be moved to act as a height adjustment. Finally, another flat washer and nut will be added to the section of each screw protruding through the base to secure the table tracks to the base.

Referring to *Figure 6-8*, M1's shaft is centered about 1 1/8" above its body. Adding 1/8" clearance from the base gives a final, installed height of the shaft center of about 1 1/4". Since the shaft is centered in the table driver, it is 3/8" above the table which, in turn, is 1/4" thick. The table, in turn, is 1/8" above the bottom of the table track (since the track is 1/8" thick). Subtracting these fractions from the 1 1/4" number leaves 1/2" clearance between the top surface of the base and bottom surface of the table track.

Figure 6-9 shows three flat washers (#13) and three nuts (#14) on a typical table track mounting screw (#12). The center nut is adjusted so the track is 1/2" above the base. The bottom flat washer and nut is added to secure the track to the base. This process is followed for each of the four screws in the two table tracks. The nuts can be readjusted later to compensate for any minor misalignment of the table stepper mounts (#42) or table stepper motor M1.

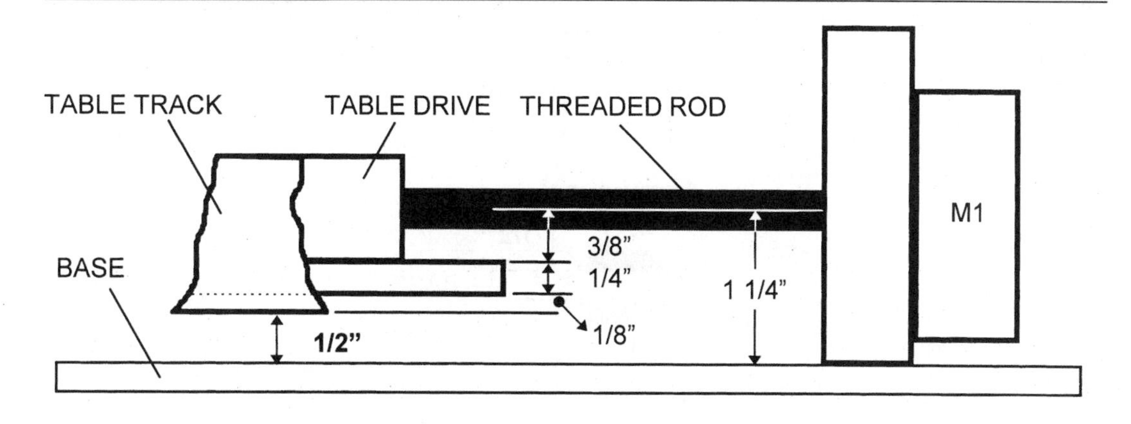

Figure 6-8*. Installment measurements of the tracks and base.*

Mount the long table track into holes 6 and 7 (*Figure 4-10*), with its vertical edge furthest away from the base's center. Similarly, mount the short table track into holes 3 and 4, with its vertical edge furthest away from the base's center. *Photo 6-2* shows the table tracks mounted to the base.

The Table Stepper Mounts

The table stepper mounts (#42) are two pieces of wood bolted to the base. They will secure the table stepper motor to the base and position it perpendicular to the table surface. To fabricate the table stepper mounts, do the following:

1. Cut two pieces of wood to final dimensions of 1" x 1 5/8" x 3/4". Arbitrarily identify one as the Left Mount, and the other as the Right Mount.
2. Drill one vertical and one horizontal 3/16" diameter holes through each mount as shown in *Figure 6-10*. Countersink the vertical hole using a 5/16" diameter drill bit to an approximate depth of 1/4".
3. Sand all surfaces smooth and finish with water based varnish or other suitable clear coating. Let dry thoroughly before proceeding.

When dry, mount the left table stepper mount in the base's hole 5 (refer to *Figure 4-10*) using a #6-32 x 2" machine screw (#9) through the mount, and into hole 5. Secure the left table stepper to the base with a #6 flat washer (#6) and a #6-32 machine nut (#11) on the bottom of the base.

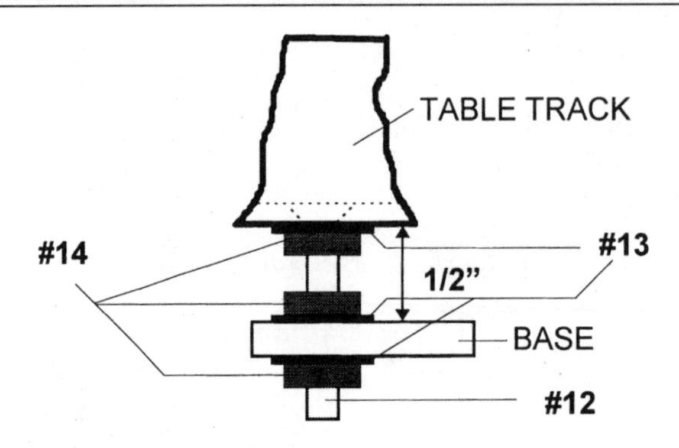

Figure 6-9*. The flat washers and nuts on a typical track mounting screw.*

Similarly, mount the right table stepper mount on the base using a #6-32 x 2" machine screw (#9) through the mount, and into hole 10. Secure the right table stepper to the base with a #6 flat washer (#6) and a #6-32 machine nut (#11) on the bottom of the base. When properly mounted, the two unpopulated holes in the mounts will be closest to each other (i.e., in the orientation shown in *Figure 6-10*). *Photo 6-3* shows the table stepper mounts mounted to the base.

Putting it All Together

Now it's time to put all the table assembly pieces together. As a preview, here's what we're going to do:

1. Prepare a threaded rod driver and attach a 1" length of plastic tubing (or heat shrink tubing) onto one end of the rod.
2. Thread the rod into the table driver.
3. Install the table surface on the table tracks.
4. Install the table guide onto the table driver.
5. Slide a tubing clamp and then the free end of the tubing (or just slide the free end of the heat shrink tubing) onto the shaft of the stepper motor and tighten the clamp (or shrink the heat shrink tubing).
6. Install the table limit switch onto one of the table stepper mounts.
7. Route the stepper and switch wires to the electronics.

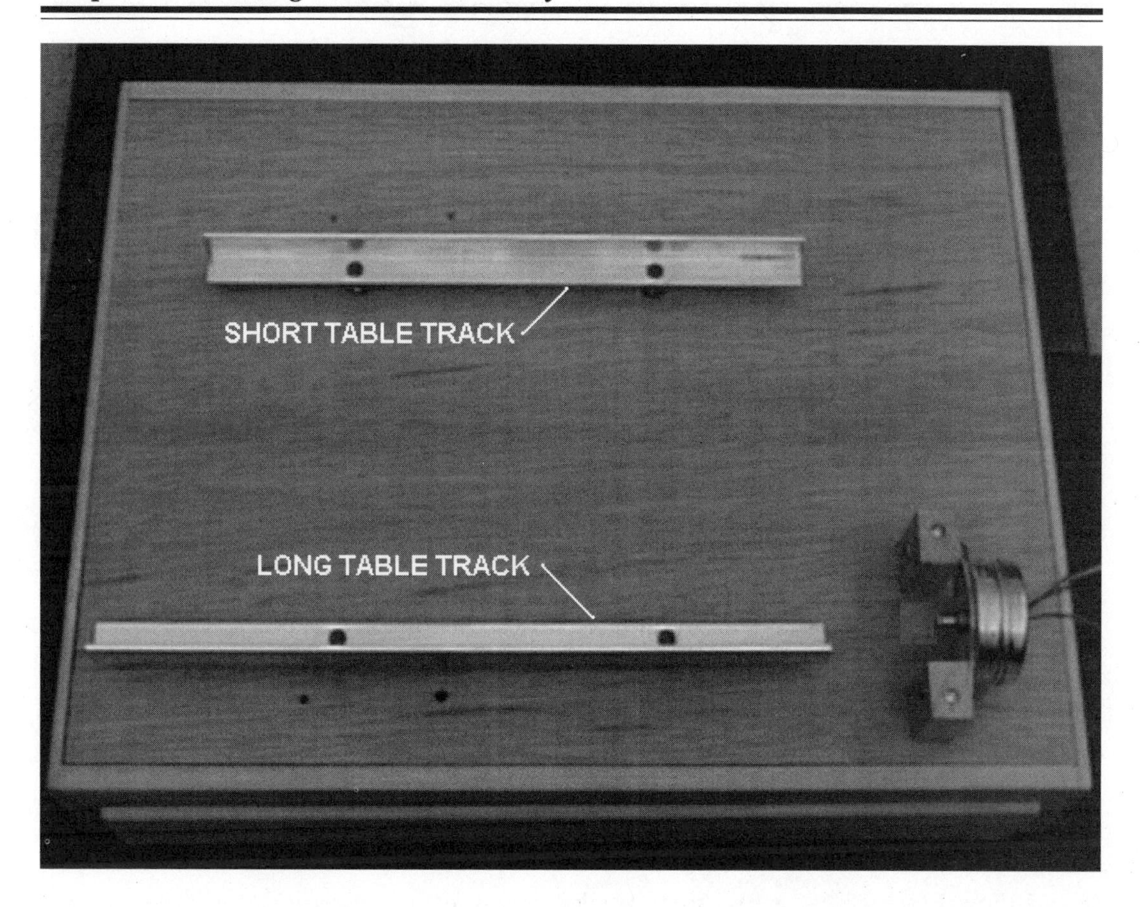

Photo 6-2. *The table tracks mounted on the base.*

Installing the Threaded Rod Driver

Obtain a 12" long threaded rod (#17). Chuck one end into a 3/8" variable speed drill. Slowly rotate the rod, noting if it wobbles from side to side as it rotates. If so, bend the rod slightly in the direction opposite the wobble. Do this as often as necessary until the rod rotates without any side-to-side wobble.

Next, expand one end of a 1" length of plastic (or heat shrink) tubing with the writing end of a ball point pen. "Thread" this end about 1/4" onto the threaded rod. Secure the plastic tubing with a tubing clamp. (If using heat shrink tubing, do not heat it at this time).

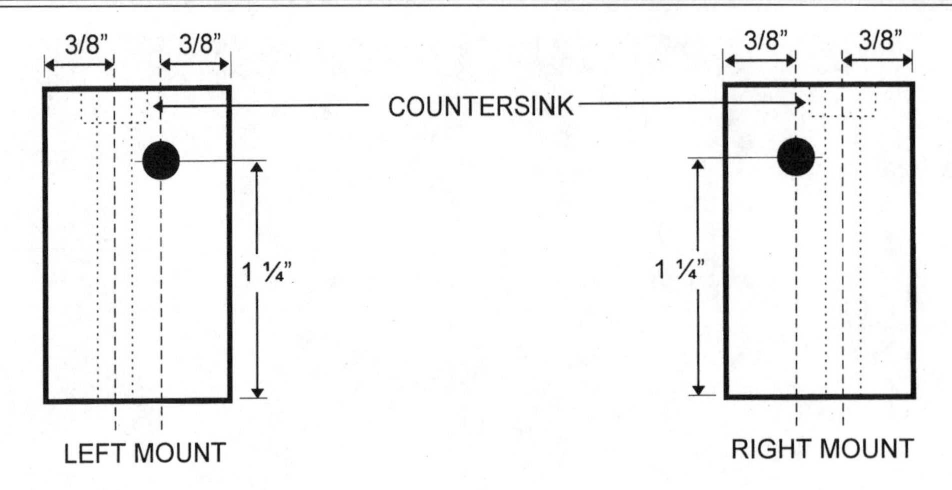

Figure 6-10*. Hole locations for the table stepper mounts.*

Next, pass the rod through the open hole in the table driver and thread it onto the nut at the driver's other end. Continue threading the rod into the table driver until it is about centered in the table driver.

Installing the Table Surface on the Table Tracks

Install the table surface on the table tracks with the tubing facing the table stepper mounts.

Installing the Table Guide

Push the table driver flush to the long table track. Place the side of the table guide with the elongated hole on top of the table driver (the other side should be on the outside of the table track). Center the table guide on the table track and mark the center point of the elongated hole on the table driver.

Momentarily remove the table guide and drill a 3/32" hole in the table driver at the mark. Reinstall the table guide and secure it to the table driver with a #4-40 x 3/4" machine screw (#3). Loosen the screw and install a business card between the outside surface of the long table track and the inside surface of the table guide. Snug the table guide against the business card and tighten the screw. Remove the business card. The installed table guide is shown in *Photo 6-3*.

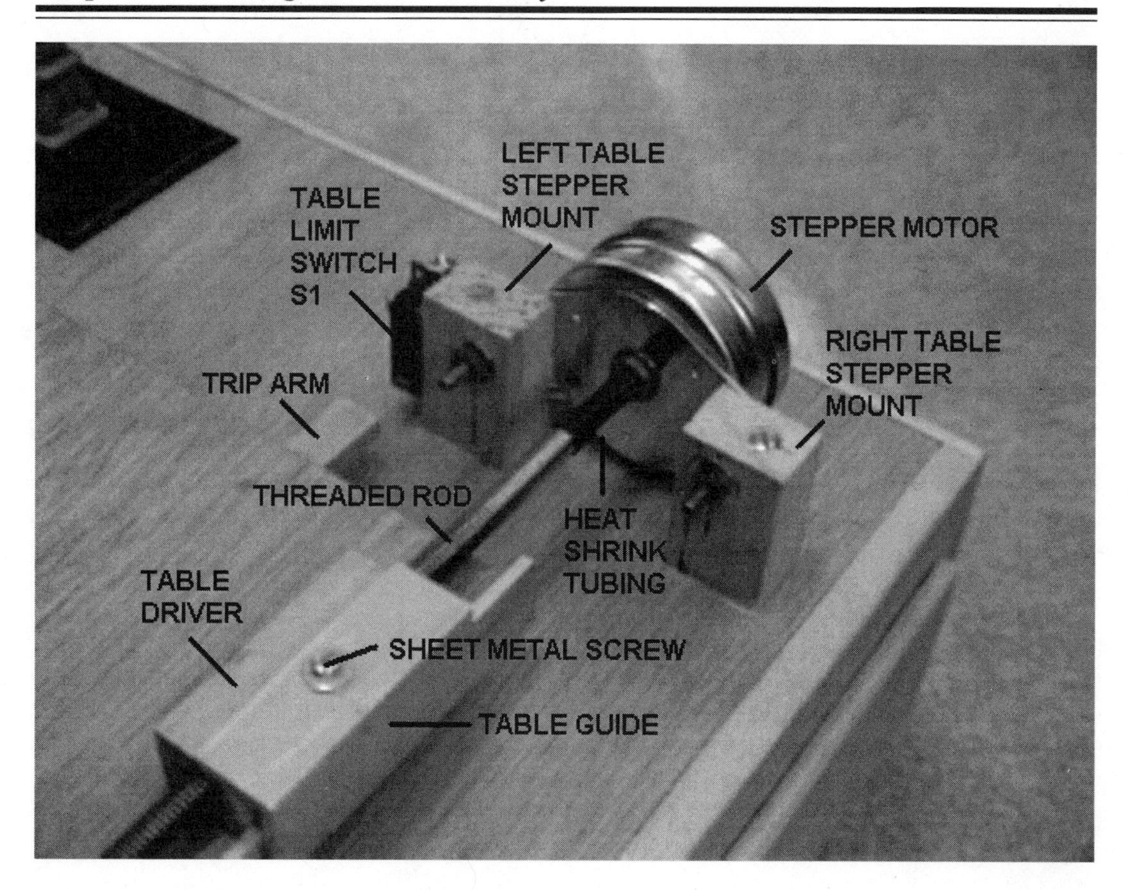

Photo 6-3*. Table stepper mounts mounted on the base.*

Installing the Table Stepper Motor

Using two #6-32 x 1 1/4" machine screws (#8), two #6 flat washers (#6), and two #6-32 nuts (#11), mount the stepper to the table stepper mounts (refer to *Photo 6-3* and *Photo 6-4*). The screw should pass through the stepper motor's mounting bracket and then through the hole in the table stepper mount. Do not secure the nut too tightly at this time, as the position of the motor will be adjusted during initial alignment.

Connecting the Threaded Rod Driver and Stepper Motor

If using plastic tubing and tubing clamp, slide the clamp (#32) over the stepper motor shaft. Then push the free end of the tubing onto the shaft. Test to ensure the end of the threaded rod and motor shaft are almost touching. Shorten the tubing as necessary. Finally, secure the clamp.

If using heat shrink tubing, push the free end of the tubing onto the shaft. Test to ensure the end of the threaded rod and motor shaft are almost touching. Shorten the tubing as necessary. Apply heat with a heat gun or high power hair dryer. As the tubing begins to shrink, rotate the assembly slightly to apply heat evenly around the tubing. Continue applying heat until the tubing has shrunk sufficiently to form a tight fit. Note that the shrink process may take some time since the threaded rod and motor shaft both work as heat sinks.

Installing the Table Limit Switch

Rotate the threaded rod until the table's trip arm is approximately flush with the front surface of the left table stepper mount. Hold a microswitch in the approximate position shown in *Photo 6-3* and *Photo 6-4* and position it so the trip arm depresses the microswitch's plunger. Mark the mounting positions through the two holes in the microswitch. Drill two 3/32" holes at the marked positions. Mount the microswitch to the left table stepper mount using two #4-40 x 3/4" machine screws (#3).

Connecting the Stepper Motor and Limit Switch

Referring to *Photo 6-4*, locate a convenient spot behind the left table stepper mount. Drill a 1/4" diameter hole and route the four wires from the table stepper motor through the hole.

Locate the two S1 wires (S1+ and S1G, *Figure 5-4*) and route them through the hole to switch S1. Solder either of the two S1 wires to the common terminal on the microswitch. Solder the other S1 wire to the normally open (NO) terminal on the microswitch.

Locate the four M1 wires (M1A, M1B, M1A* and M1B*, *Figure 5-4*) and connect them to the appropriate wires from the table stepper motor.

Initial Alignment

If everything is parallel or perpendicular where it's supposed to be, and the threaded rod and motor shaft are concentric, alignment will be very quick. Odds are, however, that there are slight misalignments of the table tracks, table driver, and table stepper

***Photo 6-4**. Mount the stepper to the table stepper mounts.*

mounts. In addition, the threaded rod is probably not totally straight, and there is probably some non-concentricity between it and the table stepper motor. Not to worry, however, since the design allows you to make various adjustments to compensate for just about any misalignment. Initial alignment will consist of the following steps:

1. Perform the alignments as described earlier.
2. Run the SPEEDSET.EXE program.
3. Connect the *Auto-XY* to your PC.
4. Power up the *Auto-XY*.
5. Test for smooth table movement.
6. Make any minor adjustments necessary to ensure the table moves smoothly and quietly throughout its total movement range.

Be prepared to invest some time in doing this initial alignment. Don't get nervous if the first time you try to move the table, it just sits there and chatters. *This is normal!* Just follow the steps below methodically and you'll have a smoothly running table assembly.

The Alignment Points

There are four separate "points" where you can make adjustments that affect the alignment. They are:

1. The table rails.
2. The table guide.
3. The table stepper mounts.
4. The stepper motor position on the table stepper mounts.

The natural urge is to adjust one thing then move another thing randomly. Don't succumb to that urge! To ensure that alignment goes smoothly, follow the alignment sequence below.

The Table Rails

The first alignment step is to make the table rails parallel to the base. Since the table moves along the table rails, any non-parallelism will be repeated in the threaded rod driver and cause the stepper motor to bind.

Referring to *Figure 6-8* and *Figure 6-9*, loosen the nuts on the bottom of the base that hold the table tracks in place. Adjust the table rail nuts resting on the base so the bottom of each rail is exactly 1/2" above the base. Check this dimension at various points along the length of each table track.

Next, you want to make the long (and short) table track parallel to the long edge of the base. Measure the distance from the side of the long table track to the edge of the base. Adjust the long table track until the distance is exactly the same at various points along the length of the track. Secure the long track in place by tightening the nuts on the bottom of the base.

Now measure between the inside surfaces of the long and short tracks. Adjust the short track until the distance is exactly the same at various points along the length of the tracks. Secure the short track in place by tightening the nuts on the bottom of the base.

Apply a light coating of light household (3-in-1 type) oil to the surfaces of the both rails. Wipe off any excess with a paper towel.

The Table Guide

Aligning the table guide consists of ensuring it is at a uniform distance from the large table track. To do this, loosen the #4-40 machine screw holding the table guide to the table drive. Install a business card between the outside surface of the long table track and the inside surface of the table guide. Snug the table guide against the business card and ensure the guide is evenly contacting the card. Tighten the screw, and then remove the business card.

The Table Stepper Mounts

Now that the previous alignments have established the table (and its threaded rod driver) parallel to the edge of the base, you need to ensure the table stepper mounts are aligned to the edge of the base as well. To do this, obtain a combination square, "speed" square, "T" square or other such device that has an exact 90 degree angle.

Loosen the screws on the bottom of the base that are holding the stepper mounts in place. Rest one edge of the square against the edge of the base and move the square until it contacts the surface of the right stepper mount. Adjust the position of both stepper mounts so they align with the edge of the square. Retighten the stepper mount mounting screws.

Using the same square, check the vertical angle between the base and the height of each stepper mount. If it is not exactly 90 degrees, re-loosen the mounting screws, install appropriately sized adjusting shims under the mount(s), and retighten the mounting screws. (Shims can be made from sections of a matchbook cover.)

The Stepper Motor

Now that all other elements have been aligned, it's time to align the stepper motor. Loosen the two nuts holding the stepper motor in place and move the stepper motor side to side while measuring the distance from the threaded rod to the long table track. When this distance is equal along the length of the rod, mark the side position of the stepper motor on the rear of the left stepper mount.

Move the stepper motor up and down while measuring the distance from the threaded rod to the table. When the distance is equal along the length of the rod, make sure the stepper is still on the side-to-side mark made previously. Then, tighten the nuts holding the stepper motor. Mark the outline of the stepper motor's mounting flange on the rear of both stepper mounts.

Testing the Alignment

Connect the *Auto-XY* to your PC's parallel port and power up the unit. Start up your PC and run the SPEEDSET.EXE program. Use the 4 and 6 keys to move the table back and forth along its full length of travel. As the table moves back and forth, drip some light household (3-in-1 type) oil onto the threaded rod near the pressed-in nut end of the table driver so the oil is dispersed on the rod and the nut in the driver. Also, apply oil along the length of the rod. This lubrication will make a noticeable improvement in the table's smooth movement. As the table reaches the end of its travel towards the stepper motor, note that as the trip arm contacts the table limit switch, movement stops.

If the table resists, skips, or chatters during travel, the first adjustment to make is the motor position. Slightly loosen the nuts holding the motor in place and run the table back and forth. If the table binds or hesitates, slightly move the motor until the table runs freely. When you are satisfied that the table is running smoothly along its complete travel length, retighten the nuts.

It is also possible that the pressed-in nut in the table driver is not perpendicular to the threaded rod. To adjust this, run the table away from the stepper motor until the threaded rod disengages from the table driver. Thread another threaded rod into the nut and bend the rod slightly to move the nut in the driver towards a perpendicular position. Place the table back on the tracks and re-thread the driver onto the threaded rod attached to the stepper motor. Perform this process as often as necessary until the nut is perpendicular and the threaded rod runs smoothly along its full length of travel.

If the rod is not concentric with the motor shaft, push the rod slightly in the direction of non-concentricity. (The tubing has enough flex to allow this pushing movement to adjust the concentricity of the threaded rod and the motor shaft.) Run the table back and forth and repeat this process as often as necessary.

If you are still not satisfied with the smoothness of movement, you can make further minor adjustments to the table drive, table stepper mounts, and table guide.

Wrapping it Up

We've covered a lot of ground in this chapter, but we still have a ways to go. In the next chapter, we'll construct the drill assembly. This part of the project should go a little faster, since we'll be using the same processes we used on the table assembly. In addition, alignment is less complicated, since the drill tracks are "aligned" as part of the construction process. Once we finish the drill assembly, it's on to our first PC board using *Auto-XY*.

CHAPTER 7
Building the Drill Assembly

Introduction

We're now in the home stretch. The last assembly to be constructed is the drill assembly. It consists of the drill "bridge," and the drill "caddy." The drill bridge, constructed from angle aluminum and wood, straddles the table at 90 degrees. It allows the drill caddy to move directly above the table in the Y direction. It also holds the drill limit switch, which is tripped when the drill caddy reaches its zero position. The drill caddy contains the drill driver and accepts the drill and drill collar subassembly. The drill stepper motor is secured to the drill bridge with two machine screws.

In this chapter, we'll fabricate the drill bridge and caddy. We'll then install the drill bridge to the base, and install the drill caddy and drill/drill collar subassembly. Then we'll interconnect the limit switch and stepper motor to the control electronics. When we've completed these steps, we'll align the drill assembly using the SPEEDSET program. As we proceed, you'll need to refer back to the tables in Chapter 3 for the dimensions and other details on the individual parts.

Critical Dimensions

As previously mentioned in Chapter 3, the drill bridge and caddy dimensions reflect a maximum cross-sectional area of the drill. If you'll be using a drill with a cross-sectional area larger than that, the drill bridge and caddy dimensions must be adjusted accordingly.

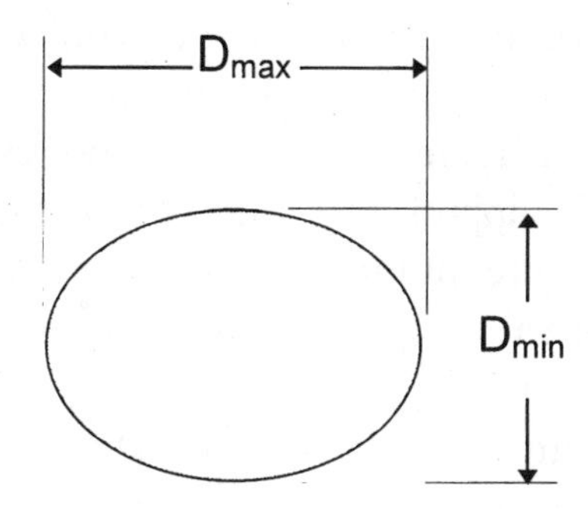

Figure 7-1*. Most drills have maximum and minimum dimensions.*

Photo 7-1*. The drill bridge resembles an upside down "U."*

Drill Bridge and Caddy

As shown in *Figure 7-1*, any drill will have a maximum (Dmax) and minimum (Dmin) dimension. (If the drill is perfectly round, Dmax will equal Dmin.) Determine both dimensions for the drill you will be using. Then substitute those values in the formulas below (refer to the specified figures to see where the resultant dimensions are used). After you've determined the appropriate dimensions, cut the material accordingly. If you are using a drill within the minimum and maximum cross-sectional area dimensions, use the parenthetical values shown in the figures (as calculated below):

Drill Bridge Spreader Length:	S = Dmax $3^{1/4}$" (*Figure 7-3*)
Drill Caddy Length:	C = Dmax + 3" (*Figure 7-10*)
Drill Caddy Side Stretcher Length:	SS = Dmax + $2^{1/4}$" (*Figure 7-9*)
Drill Driver (and Drill End Piece) Width:	W = Dmin + 1/8" (*Figure 7-8*)

So, for the specified drill with Dmax = $2^{1/4}$", and Dmin = $1^{7/8}$":

S = Dmax + $3^{1/4}$" = $2^{1/4}$" + $3^{1/4}$" = $5^{1/2}$"
C = Dmax + 3" = $2^{1/4}$" + 3" = $5^{1/4}$"
SS = Dmax + $2^{1/4}$" = $2^{1/4}$" + $2^{1/4}$" = $4^{1/2}$"
W = Dmin + 1/8" = 1 7/8" + 1/8" = 2"

Drill Motor Mount

The drill motor mount dimensions are based on the use of the specified stepper motor. If another motor is used, the two mounting hole positions must be adjusted accordingly. Referring to *Figure 7-5*, determine the distance between mounting holes on the stepper being used (calling that distance D). Then, determine the value of M using the formula:

$$M = (D/2) - 1"$$

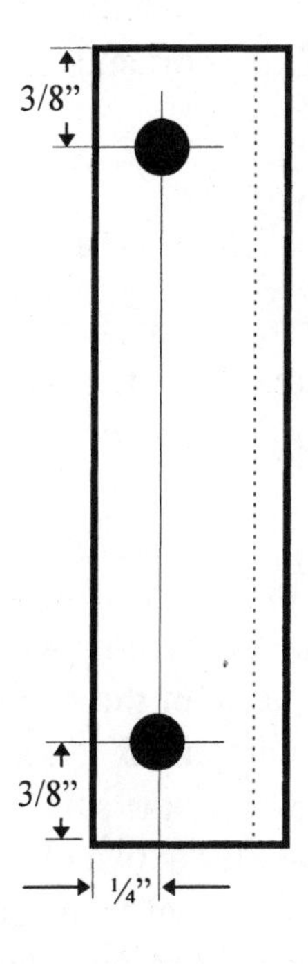

***Figure 7-2**. Drill holes in the #35 pieces as shown.*

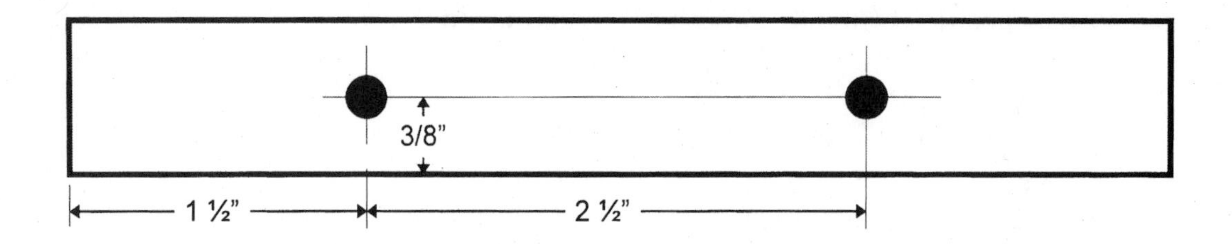

Figure 7-3. Drill two through-holes in two of the #45 pieces as shown.

So, for the specified stepper motor with D = 2.6" (or about $2^{5/8}$"), M = 1.3" - 1" = .3" (or about 13/16"). Since the diameter of the mounting holes are larger than the mounting screws (to allow for some adjustment), you can round off the value of M to the nearest sixteenth of an inch.

The Drill Bridge

As shown in *Photo 7-1*, the drill bridge resembles an elongated, upside-down "U." It rests on the base and allows the drill to move at 90 degrees to the table movement. The drill bridge uses the same type driver, threaded rod, and tubing linkage as the table. The drill bridge consists of two legs, the short and long drill caddy tracks, and the drill motor mount track.

The Drill Bridge Legs

The two drill bridge legs are similar, each consisting of two vertical aluminum angle pieces and two horizontal wood spreaders. On one leg, the top spreader is a 3/4" square wood piece. On the other it is a $1^{1/4}$" wide by 3/4" thick wood piece. To fabricate the drill legs, do the following:

1. Cut four pieces of angle aluminum 3" long (#35), making sure all four pieces are exactly the same length. These pieces will now be known as verticals.
2. Drill two 3/16" diameter holes in each of the #35 pieces as shown in *Figure 7-2*.
3. Cut three pieces of wood 3/4" square and $5^{1/2}$" long (as defined in *Figure 7-4*), making sure all three pieces are exactly the same length. These are item #45.
4. Drill two 3/16" diameter through-holes in two of the #45 pieces as shown in *Figure 7-3*. These two pieces will now be known as the bottom spreaders. The remaining piece will now be known as the small top spreader.

5. Cut one piece of wood $5^{1/2}$" long by $1^{1/2}$" wide by 3/4" thick (#46), making sure it is exactly the same length as the other three wood spreaders. This piece will now be known as the large top spreader.
6. Assemble two verticals, a bottom spreader and the small top spreader as shown in *Figure 7-4*. Mark the verticals' hole centers in the spreaders and predrill with a 3/32" drill bit. Secure with four #8 sheet metal screws (#2). Check to make sure the resultant assembly is square. Adjust as necessary.
7. Assemble the remaining two verticals, bottom spreader, and large top spreader as shown in *Figure 7-4*. Note that the large top spreader is positioned so its 3/4" thickness is vertical. Mark the verticals' hole centers in the spreaders and predrill with a 3/32" drill bit. Secure with four #8 sheet metal screws (#2). Check to make sure the resultant assembly is square. Adjust as necessary.

Completing the Drill Bridge

You'll now fabricate the long drill caddy track, short drill caddy track, and drill motor mount track. Then you'll assemble them to the drill bridge legs to complete the drill bridge:

1. Cut a drill motor mount track (#36), short drill caddy track (#37), and long drill caddy track (#38) from angle aluminum, as identified in *Table 3-3*.
2. Drill the four 3/16" diameter holes and cut a notch in the drill motor mount track as shown in *Figure 7-5a* (side view) and *Figure 7-5b* (top view).
3. Drill two 3/16" diameter holes in each end of the short and long drill caddy tracks as shown in *Figure 7-6*.
4. Referring to *Figure 7-7*, install the long drill caddy track onto the top of the two drill bridge legs as shown. Predrill the mounting holes and secure with two #8 sheet metal screws (#2).
5. Position the short drill caddy track as shown in *Figure 7-7*. Check for squaring. After any necessary adjustment, predrill two mounting holes and secure with two #8 sheet metal screws (#2).
6. Position the drill motor mount track as shown in *Figure 7-7*. Note that the inside edge of the notch aligns with the inside surface of the short drill caddy track. Check for squaring. After any necessary adjustment, predrill two mounting holes and secure with two #8 sheet metal screws (#2).
7. Referring to *Photo 7-1*, mount the drill bridge on the base using four #6-32 x $1^{1/4}$" machine screws (#8), four #6 flat washers (#6), and four #6-32 nuts (#11).

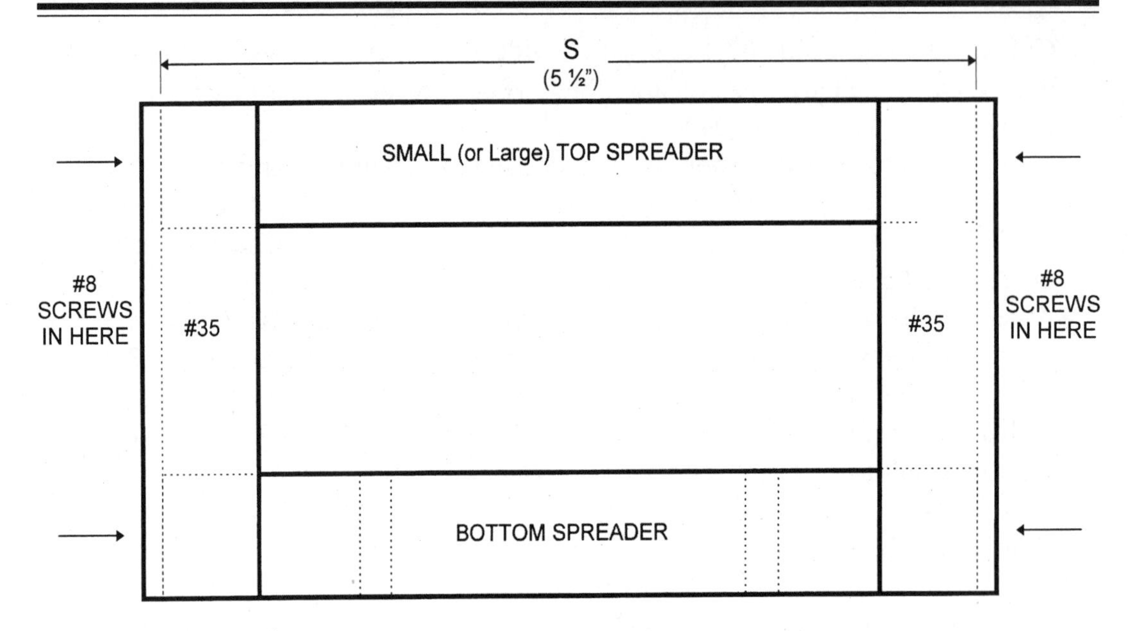

Figure 7-4. *Assemble two verticals, a bottom spreader and the small top spreader as shown.*

The Drill Caddy

The drill caddy is a critical assembly. The precision of the holes that you will drill with the *Auto-XY* depends on the ability of the drill caddy to move the drill smoothly, and the ability of the drill to move exactly vertically. The drill caddy consists of a drill driver (#48), a drill caddy end piece (#49), two metal drill caddy side stretchers (#33), two modified lag bolts (#15), four mounting nuts (#16), one compression spring (#22), and a drill collar (fabricated from miscellaneous pieces of lumber - #60).

Step 1

In this step, we'll fabricate a drill driver (#48), and the drill caddy end piece (#49):

1. Cut a piece of wood to dimensions 3/4" thick by 1" long by 2" wide (refer to *Figure 7-8a*). This is the drill caddy end piece, item #49. Machine the two notches in the end piece as shown in *Figure 7-8a*.
2. Cut a piece of wood to the dimensions shown for the drill driver (#48) in *Figure 7-8b*.

3. Referring to end view of *Figure 7-8b*, locate the drilling center point on each end of the drill driver. Using a center punch, mark each center.
4. Using a drill press with a 3/32" bit and some form of guide, hold the driver blank on its end and parallel to the drill bit. Drill the hole about halfway through the blank. Turn the blank end-for-end and drill the other center-punched hole.
5. Using successively larger bits, continue this process until you have a through-hole that is slightly smaller than the outside dimensions of the drive nut you'll be using. (For the specified 5/16" threaded rod, that final drill bit size should be 1/2".)
6. Place a nut on the end of the blank (as shown in the end view of *Figure 7-8b*), approximately centered in the hole. Temporarily hold it there with masking tape.
7. Insert the assembly into a vise and *slowly* press the nut into the hole.
8. Drill the 9/64" through-hole and 1/2" countersink as shown in *Figure 7-8b*.

Step 2

In this step, we'll fabricate the drill caddy side stretchers, and assemble the driver, end piece, and stretchers:

1. Cut two pieces of flat aluminum stock (#33) to the length shown in *Figure 7-9*. These are the two side stretchers.
2. Drill four 1/8" holes in each side stretcher as shown in *Figure 7-9*.
3. Position the driver, end piece, and two side stretchers as shown in *Figure 7-10*. Mark the center of the eight holes, and drill a 3/32" pilot hole at each hole's center. Secure the pieces together with eight #4-40 x 3/4" machine screws (#3).

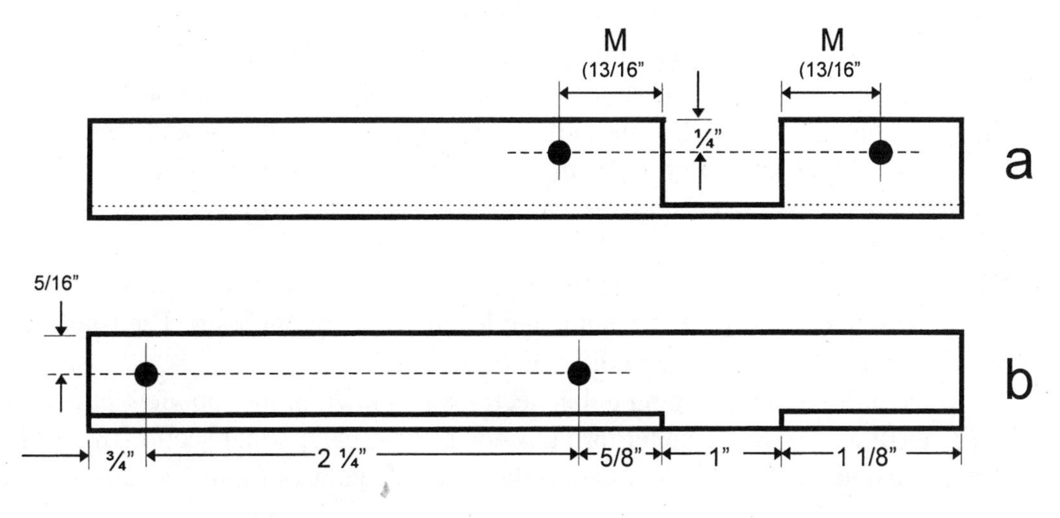

***Figure 7-5**. Notch and drill holes in the drill motor mount track as shown.*

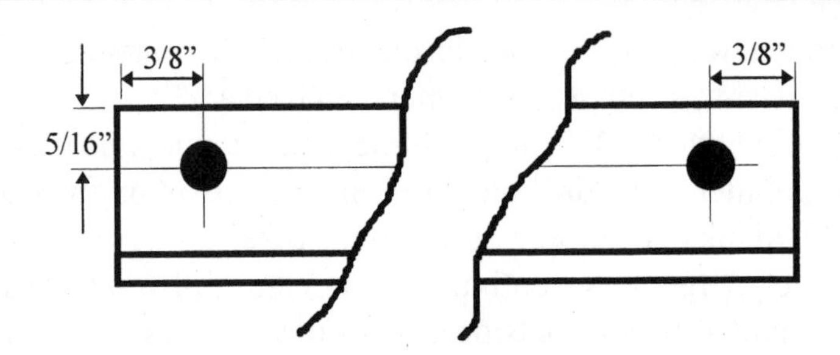

Figure 7-6*. Drill holes in the long and short drill caddy tracks as shown.*

Step 3

In this step, we'll fabricate the drill collar from miscellaneous dimensional lumber (#60), two 8-32 x 4" machine screws (#10), and two nuts (#12):

1. Referring to *Figure 7-11a*, determine the perimeter profile of the drill you will be using. (The figure shows a drill as viewed from the top). Form a pattern that is 1/8" shorter than one half the width of the drill, and has at least a 3/4" border all around the perimeter. The pattern for the drill specified in Chapter 3 is shown full size in *Figure 7-11b*.
2. Cut a collar section from 3/4" thick miscellaneous lumber (#60) using the pattern as a guide. Try the collar section on the drill, ensuring a tight fit. This piece is now called the clamping collar section.
3. Create a second collar section identical to the first.
4. Cut a section off that second collar as shown in *Figure 7-11c*.
5. Cut a 3/4" thick piece of hardwood (if available, or pine if not) to the same width as the collar, and $2^{1/2}$" high. Locate the center point of the thickness and width and drill a 1/4" diameter through-hole (making sure the hole is as straight as possible). Trim off the excess material as shown in *Figure 7-11d*. This piece will now be called the drill pusher.
6. Referring to *Figure 7-11e*, secure the modified collar piece and the drill pusher using carpenter's glue (#59). Clamp and let set for about an hour. The finished piece is now called the pusher collar section.
7. Align the pusher and clamping collar sections as shown in the top view of *Figure 7-12*. Drill two 3/16" diameter through-holes in the each collar section, making sure these holes are as straight as possible. The side view of *Figure 7-12* shows the positioning of the two through-holes.

8. Form a wedge-shaped notch in the clamping drill collar as shown in *Figure 7-13*. The narrowest part of the wedge is on the top surface of the collar, and is just large enough for the unthreaded portion of the 1/4" bolt to slide easily, but not rock side to side.
9. Assemble the two collar sections to the drill using two #8-32 x 4" machine screws (#10) and nuts (#14). Position the collar as shown in *Figure 7-14*, with the collar's bottom surface about $2^{3/4}$" above the bottom of the drill's chuck. Check to ensure that the collar is perpendicular to the body of the drill (such that if the collar were rested on a horizontal surface, the body of the drill would be perfectly vertical).

Step 4

In this step, we'll modify two 1/4" x 4" long lag bolts (#15) to serve as guides, assemble the guides to the drill caddy with four 1/4" bolts and washers (#16), and install a spring (#22) on one of the drill guides:

1. Cut the head off of both lag bolts to obtain two headless bolts about $3^{1/2}$" in length. File the cut end on the bolts so the ends are smooth.

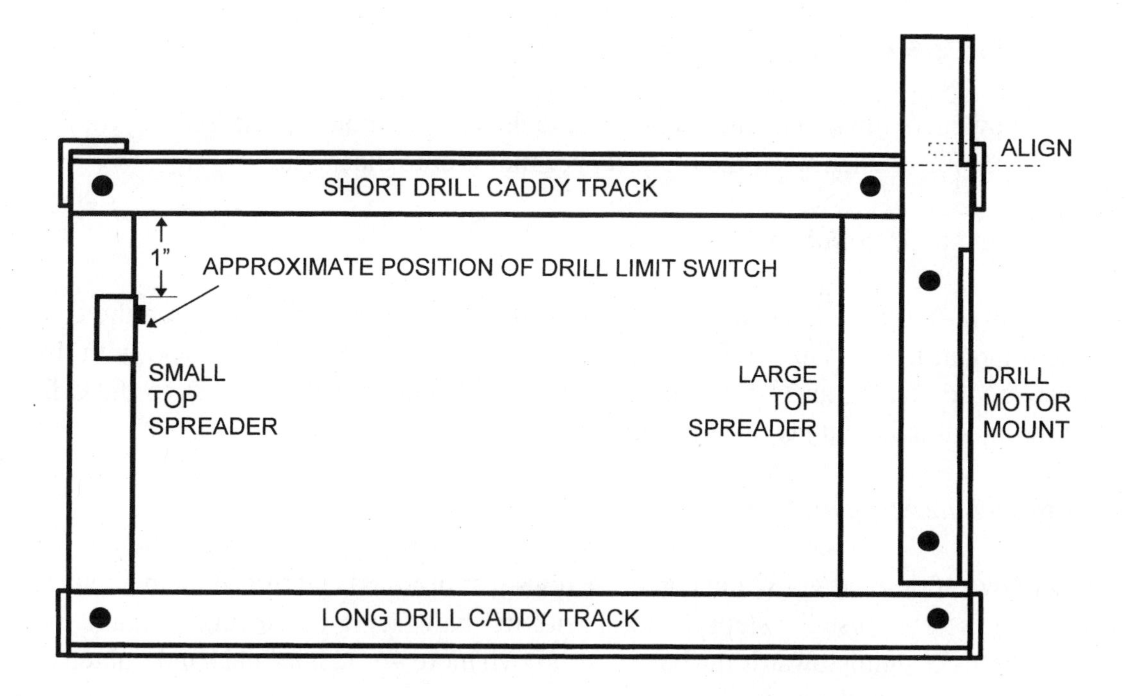

Figure 7-7*. Install the long drill caddy track onto the top of the two drill legs as shown.*

2. Mount one bolt in the countersunk hole in the drill driver as shown in *Figure 7-15*. Secure with a nut on the top and bottom of the driver (the nut on the bottom of the driver sits in the countersink).
3. Mount the other bolt approximately centered in the slot of the drill caddy end piece (also shown in *Figure 7-15*). Install a washer (#16) on the bolt as shown, and lightly secure the bolt with two more 1/4" nuts (#16).
4. Slip a washer (#16), spring (#22), and another washer over the bolt on the drill driver as shown in *Figure 7-15*.

Step 5

In this last step, we'll align the bolt in the drill caddy end piece for smooth operation:

1. Slide the drill and drill collar subassembly onto the bolts, with the drill pusher going over the bolt in the drill driver.
2. Adjust the position of the bolt in the drill caddy end piece so it is approximately centered in the notch in the clamping collar piece. Tighten the nuts holding this bolt.

Installing the Drill Stepper Motor

We'll now install two machine screws on the drill motor mount track (see *Figure 7-7* and *Photo 7-2*). Then we'll mount a stepper motor onto those screws.

Installing the Mounting Screws

Install two #6-32 x 2" machine screws (#9) into the two mounting holes on the drill motor mount track. The screw heads should rest on the inside of the track, with the screw bodies facing away from the drill bridge. Secure the two screws to the drill motor mount track with two #6-32 nuts (#11).

Installing the Stepper Motor

Install two additional #6-32 nuts on the screws just installed and thread them about 1/4" on the screw. Install a stepper motor with its shaft facing the drill bridge, and the motor leads pointing towards the base. Install two more #6-32 nuts and hand tighten to temporarily secure the motor into place.

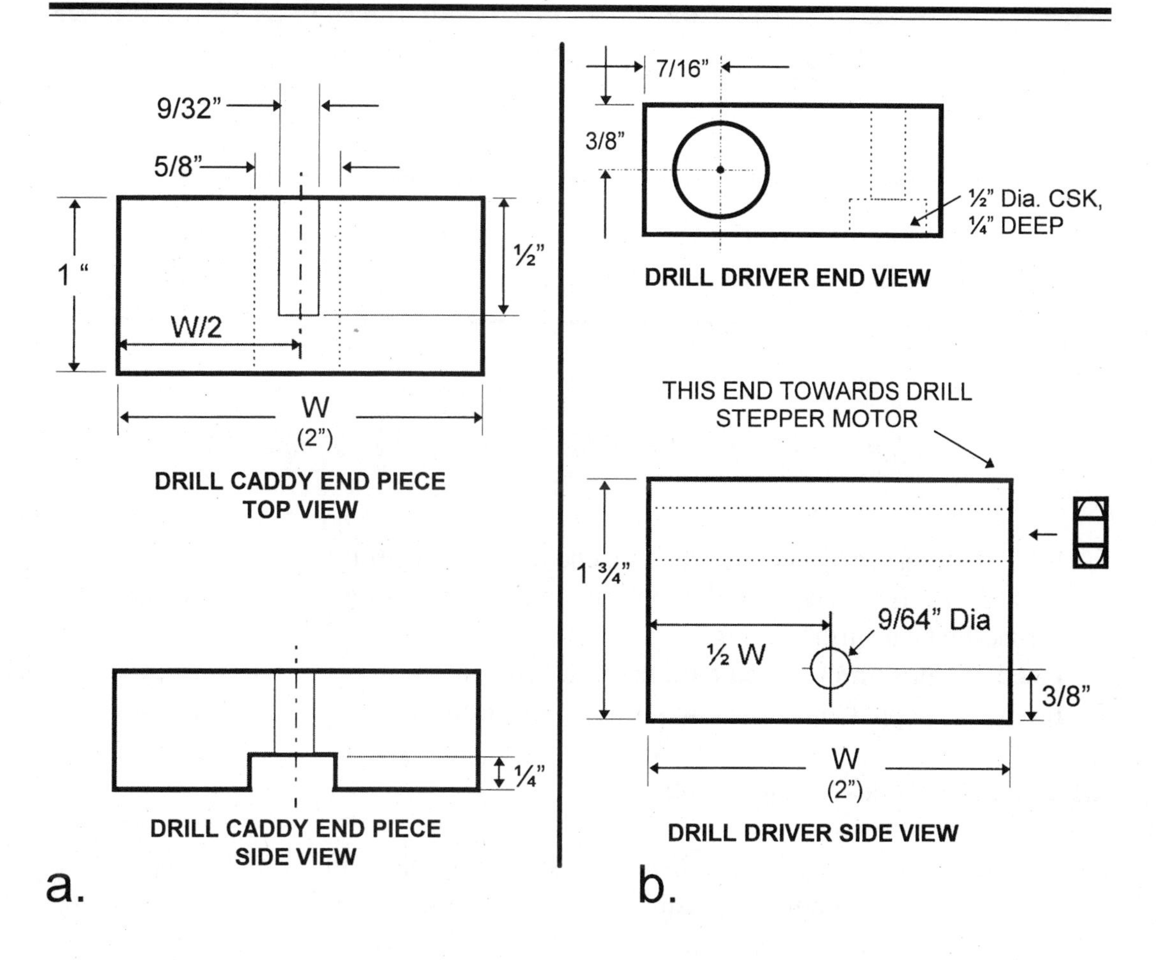

***Figure 7-8.** Cut and drill the pieces for the drill driver and drill caddy end pieces as shown.*

Putting it All Together

Now it's time to put the drill pieces together. As a preview, here is what we're going to do:

1. Prepare a threaded rod driver and attach a 1" length of plastic tubing (or heat shrink tubing) onto one end of the rod.
2. Thread the rod into the drill caddy.
3. Install the drill caddy on the drill bridge.
4. Install the drill guide onto the drill caddy.

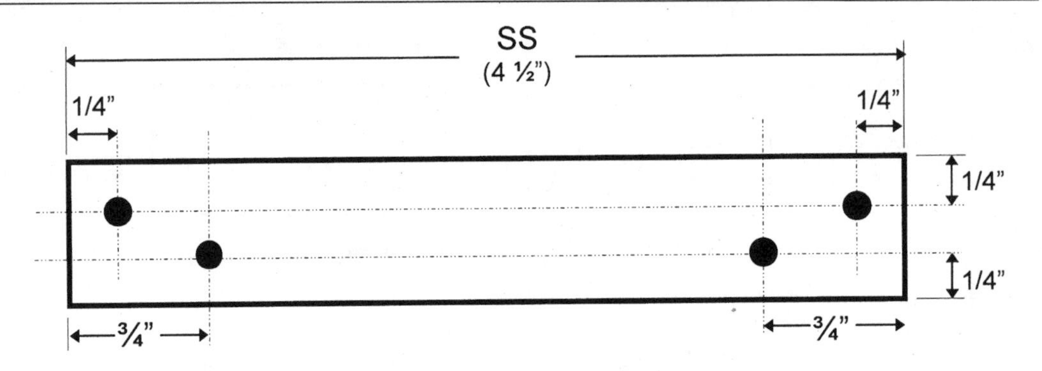

Figure 7-9. *Cut and drill the side stretchers as shown.*

5. Slide a tubing clamp and then the free end of the tubing (or just slide the free end of the heat shrink tubing) onto the shaft of the stepper motor and tighten the clamp (or shrink the heat shrink tubing).
6. Install the drill limit switch onto the drill bridge.
7. Route the stepper and switch wires to the electronics.

Installing the Threaded Rod Driver

Obtain a 12" long threaded rod (#17). Chuck one end into a 3/8" variable speed drill. Slowly rotate the rod, noting if it wobbles from side to side as it rotates. If so, bend the rod slightly in the direction opposite the wobble. Do this as often as necessary until the rod rotates without any side-to-side wobble.

Next, expand one end of a 1" length of plastic (or heat shrink) tubing with the writing end of a ball point pen. "Thread" this end about 1/4" onto the threaded rod. Secure the plastic tubing with a tubing clamp. (If using heat shrink tubing, do not heat it at this time). *Photo 7-2* shows this installation using poly tubing and a clamp.

Next, thread the free end of the rod onto the nut in the drill driver. Continue threading the rod until it is about centered in the drill driver.

Installing the Drill Caddy on the Drill Bridge

Place the drill caddy on the tracks of the drill bridge, with the tubing facing the drill stepper motor.

Installing the Drill Guide

Push the drill caddy flush to the short drill caddy track. Place the side of the drill guide (fabricated in Chapter 6) with the elongated hole on top of the drill driver (the other side should be on the outside of the track). Center the drill guide on the drill driver and mark the center point of the elongated hole on the drill driver.

Momentarily remove the drill guide and drill a 3/32" hole in the drill driver at the mark. Reinstall the drill guide and secure it to the drill driver with a #4-40 x 3/4" machine screw (#3). Loosen the screw and install a business card between the outside surface of the track and the inside surface of the drill guide. Snug the drill guide against the business card and tighten the screw. Remove the business card.

Connecting the Threaded Rod Driver and Stepper Motor

If using plastic tubing and tubing clamps, slide a clamp (#32) over the stepper motor shaft. Then push the free end of the tubing onto the shaft. Test to ensure the end of the threaded rod and motor shaft are almost touching. Shorten the tubing as necessary. Finally, secure the clamp.

If you're using heat shrink tubing, push the free end of the tubing onto the shaft. Test to ensure the end of the threaded rod and motor shaft are almost touching. Shorten the tubing as necessary. Apply heat with a heat gun or high power hair dryer. As the tubing begins to shrink, rotate the assembly slightly to apply heat evenly around the tubing. Continue applying heat until the tubing has shrunk sufficiently to form a tight fit. Note that the shrink process may take some time since the threaded rod and motor shaft both work as heat sinks.

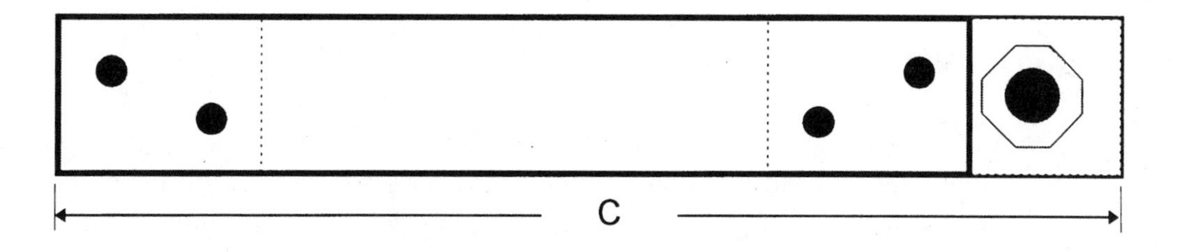

Figure 7-10*. Position the driver, end piece and two side stretchers as shown.*

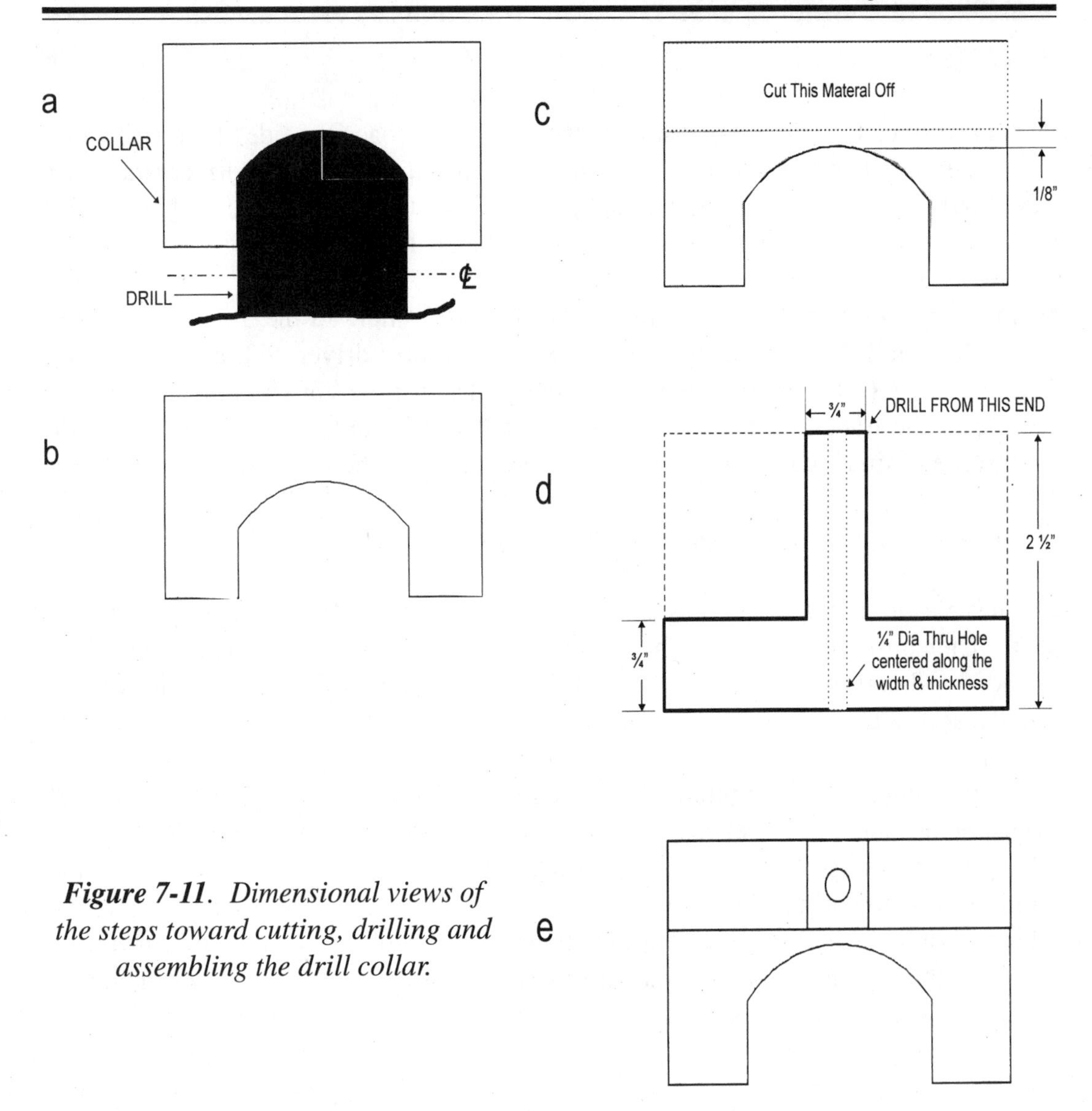

Figure 7-11. Dimensional views of the steps toward cutting, drilling and assembling the drill collar.

Installing the Drill Limit Switch S2

Referring to *Figure 7-7*, rotate the threaded rod until the drill caddy is almost flush with the short top spreader. Hold a microswitch in the approximate position shown in *Figure 7-7* and position it so the side of the drill caddy depresses the microswitch's plunger. Mark the mounting positions through the two holes in the microswitch. Drill two 3/32" holes at the marked positions. Mount the microswitch to the short top spreader using two #4-40 x 3/4" machine screws (#3).

Connecting the Stepper Motor and Limit Switch

Referring to *Photo 7-3*, locate a convenient spot below the drill stepper motor. Drill a 1/8" diameter hole and route the four wires from the table stepper motor through the hole.

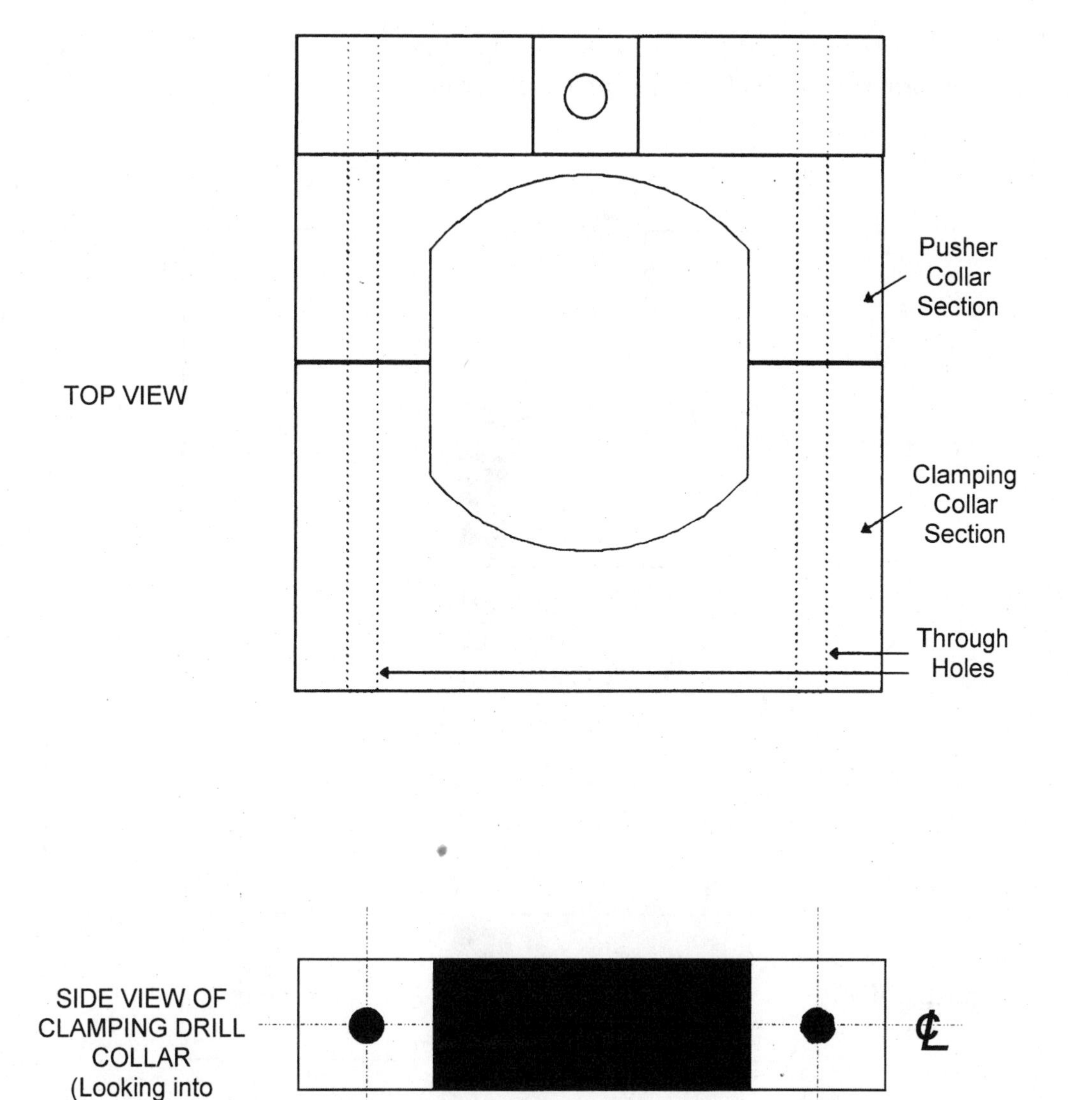

Figure 7-12*. Align the pusher and clamping collar sections as shown.*

Locate another convenient spot below the limit switch. Drill a 1/8" diameter hole at this spot. Locate the two S2 wires (S2+ and S2G, *Figure 5-4*) and route them through the hole to switch S2. Solder either of the two S2 wires to the common terminal on the microswitch. Solder the other S2 wire to the normally open (NO) terminal on the microswitch.

Locate the four M2 wires (M2A, M2B, M2A* and M2B*, *Figure 5-4*) and connect them to the appropriate wires from the table stepper motor.

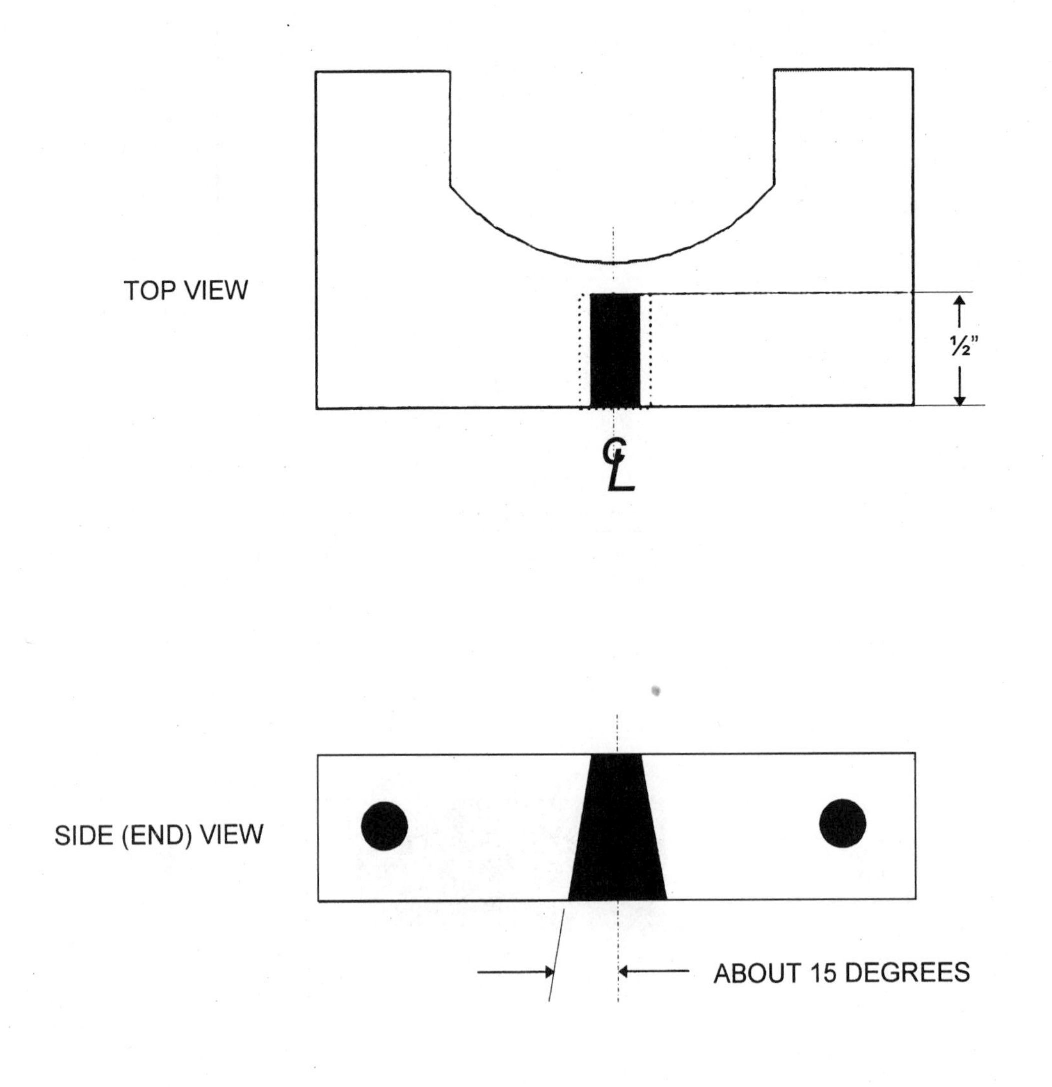

Figure 7-13*. Form a wedge-shaped notch in the clamping drill collar as shown.*

Initial Alignment

If everything is parallel or perpendicular to where it's supposed to be, and the threaded rod and motor shaft are concentric, alignment will be very quick. It is more likely, however, that the stepper motor will need to be adjusted so it positions the threaded rod parallel to the short drill caddy track. In addition, the threaded rod is probably not

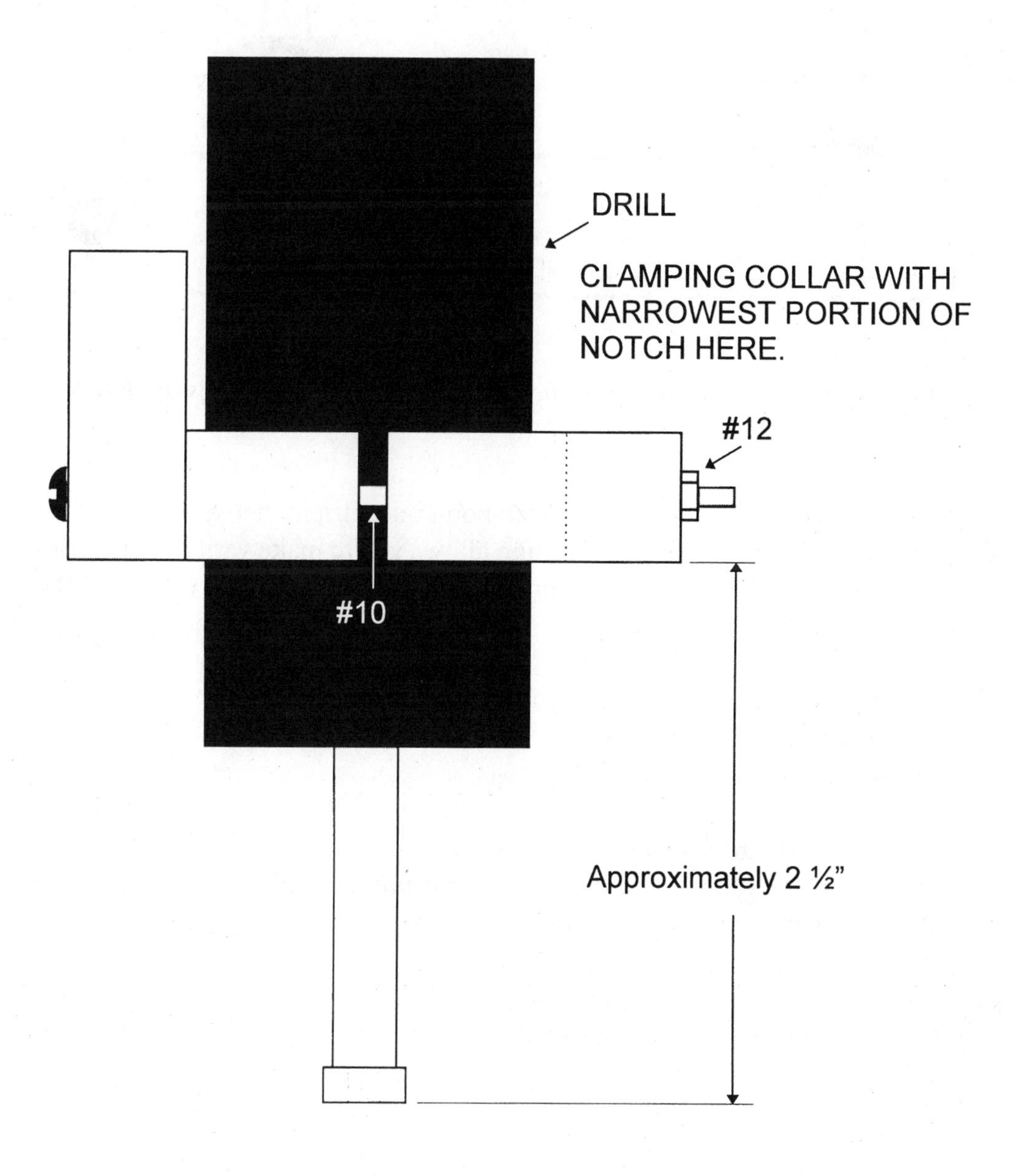

***Figure 7-14**. Position the collar as shown.*

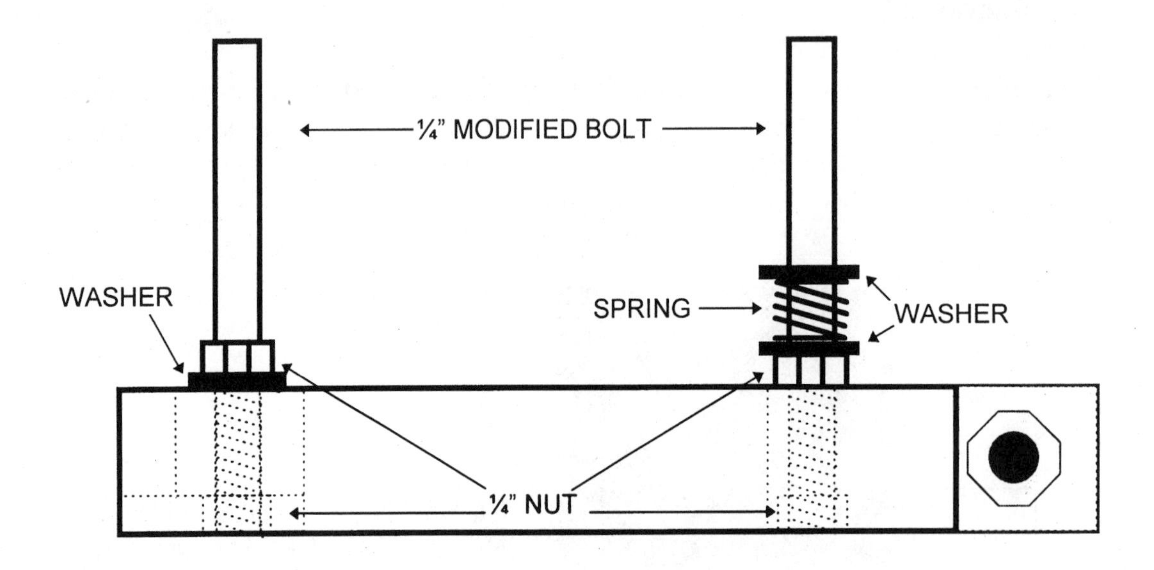

Figure 7-15*. Mount the bolts and washers for the drill guides as shown.*

totally straight, and there is probably some non-concentricity between it and the table stepper motor. As with the table, the design allows you to make various adjustments to compensate for just about any misalignment. Initial alignment will consist of the following steps.

1. Perform the alignments in described the previous paragraphs.
2. Run the SPEEDSET.EXE program.
3. Connect the *Auto-XY* to your PC.
4. Power up the *Auto-XY*.
5. Test for smooth table movement.
6. Make any minor adjustments necessary to ensure the drill caddy moves smoothly and quietly throughout its total movement range.

As with the table alignment, be prepared to invest some time in doing this initial alignment. Again, follow the steps below methodically and you'll have a smoothly running drill caddy.

Aligning the Drill Caddy

There are three separate "points" where you can make adjustments that affect the alignment. They are:

1. The drill guide.
2. The drill stepper machine screw mounts.
3. The stepper motor position on the mounts.

This alignment is somewhat more simple than that for the table. Let's begin with the drill guide.

Photo 7-2*. Install the threaded rod driver on the drill motor mount track as shown.*

Photo 7-3*. Locate a good spot for the stepper wires below the drill stepper motor.*

The Drill Guide

Aligning the drill guide consists of ensuring it is at a uniform distance from the short drill caddy track. To do this, loosen the #4-40 machine screw holding the drill guide to the drill drive. Install a business card between the outside surface of the track and the inside surface of the guide. Snug the guide against the business card and ensure the guide is evenly contacting the card. Tighten the screw, and then remove the business card.

The Drill Stepper Screw Mounts

Position the stepper motor so the mounting screws are centered in the motor's mounting holes and hand tighten the nuts to hold the motor in position. Now, adjust the position of the mounting screws so the threaded rod attached to the motor is parallel (both horizontally and vertically) with the short drill caddy track. If you cannot quite obtain the parallel position, get the rod as close as possible. Tighten down the nuts holding the screws to the drill motor mount track. *Photo 7-4* shows an end view of the drill caddy with the drill/drill collar installed.

The Stepper Motor

Now that all other elements have been aligned, it's time to align the stepper motor. Loosen the two nuts holding the stepper motor in place and move the stepper motor to make the threaded rod as parallel to the short drill caddy track as possible. Tighten the nuts to secure the motor in place.

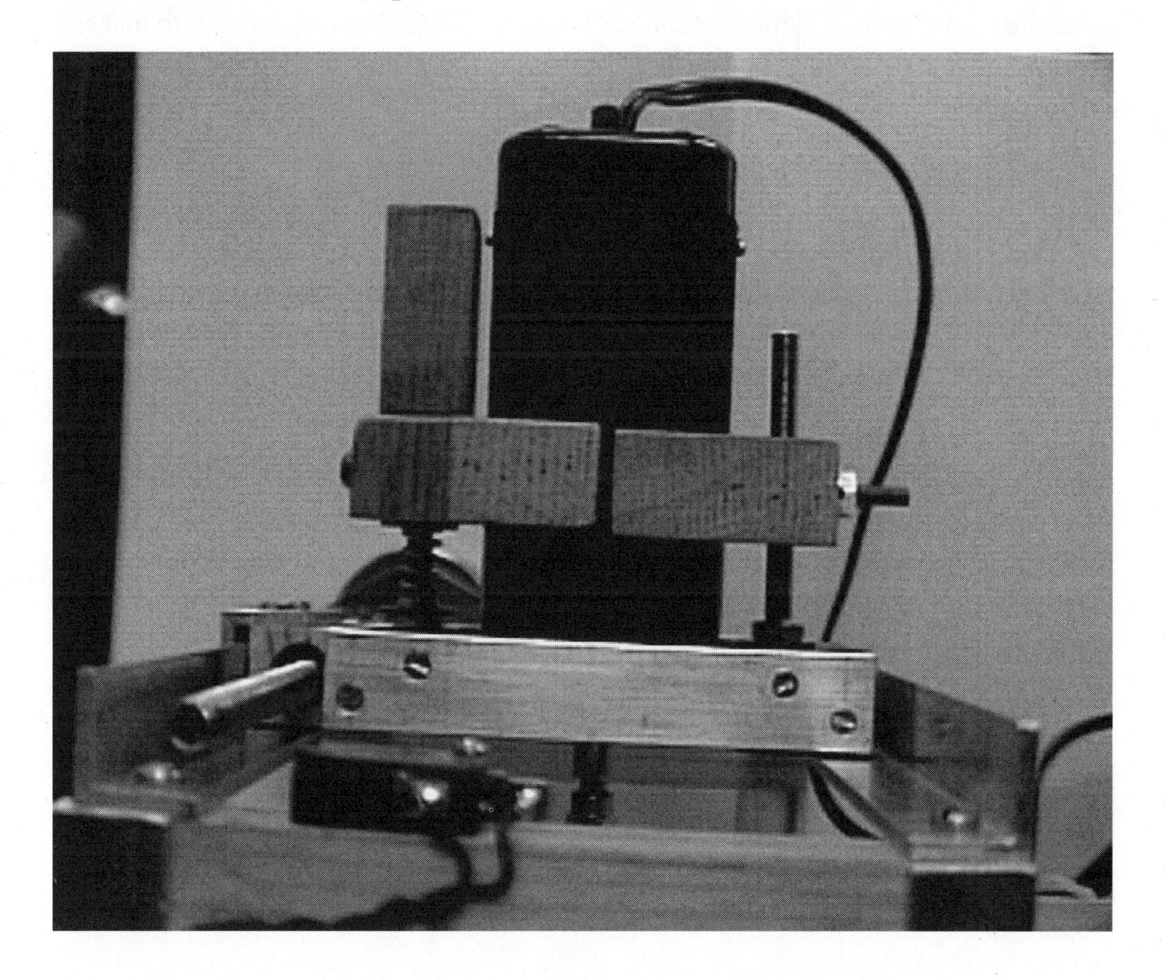

***Photo 7-4**. End view of the drill caddy with the drill/drill collar installed.*

Testing the Alignment

Connect the *Auto-XY* to your PC's parallel port and power up the unit. Start up your PC and run the SPEEDSET.EXE program. Use the 2 and 8 keys to move the drill forward and back along its full length of travel. As the table moves forward and back, drip some light household (3-in-1 type) oil onto the threaded rod near the pressed-in nut end of the drill driver so the oil is dispersed on the rod and the nut in the driver. Also, apply oil along the length of the rod. This lubrication will make a noticeable improvement in the drill's smooth movement. As the drill reaches the end of its travel away from the stepper motor, note that as the side of the drill caddy contacts the drill limit switch, movement stops.

If the drill caddy resists, skips, or chatters during travel, the first adjustment to make is the motor position. Slightly loosen the nuts holding the motor in place and run the drill forward and back. If the drill binds or hesitates, slightly move the motor until the drill caddy runs freely. When you are satisfied that the drill caddy is running smoothly along its complete travel length, retighten the nuts.

It is also possible that the pressed-in nut in the drill driver is not perpendicular to the threaded rod. To adjust this, run the drill caddy away from the stepper motor until the threaded rod disengages from the drill driver. Thread another threaded rod into the nut and bend the rod slightly to move the nut in the driver towards a perpendicular position. Place the nut in the drill driver against the threaded rod and re-thread the rod onto the driver. Perform this process as often as necessary until the nut is perpendicular and the threaded rod runs smoothly along its full length of travel.

If the rod is not concentric with the motor shaft, push the rod slightly in the direction of non-concentricity. (The tubing has enough flex to allow this pushing movement to adjust the concentricity of the threaded rod and the motor shaft.) Run the drill caddy forward and back and repeat this process as often as necessary.

If you are still not satisfied with the smoothness of movement, you can make further minor adjustments to the drill guide or drill stepper screw mounts. In fact, you may try bending the motor up or down if it seems the motor shaft is not on the same horizontal plane as the threaded rod.

Wrapping it Up

It's been a long haul, but you now have a fully functioning unit! In the next chapter, we'll install the electronics and power supply in the case. We'll also construct a momentary power switch for the drill and install a duplex outlet that will allow that power switch to control the drill. The final construction step will be to mark the zero positions on the table. Then we'll be on to fabricating our first PC board using *Auto-XY*!

CHAPTER 8
Wrapping Up Construction

Introduction

In this last construction chapter, we'll install the control electronics PC board, power supply PC board, and transformer into the case. Then we'll install a modified AC duplex outlet in the rectangular hole in the base. We'll wire this up to the AC power, and construct a switch box that will allow us to momentarily power the drill. Having the drill run continuously during drilling is unnecessary, and can cause unwanted vibration. The switch box will switch power to the drill when the box's normally open switch is depressed.

With the hardware completed, we'll adjust the parallel port, speed, and distance parameters in the setup file (DRLSETUP.DAT). Finally, we'll round out this chapter by locating the X-Y origin (X = 0 and Y = 0), and creating X and Y axis lines on the table. The axis lines and origin will ensure that any PC board mounted to the table will be aligned with the X and Y movement of *Auto-XY*.

Mounting the Electronics and Transformer in the Case

We now want to mount the control electronics PC board to the bottom of the case, and route the control cable (terminated with P1) out the rear of the case. We also want to route an AC power cord out the rear of the case, so as we remove the case rear to machine it for the control cable, we'll also machine it for the AC power cord.

Removing and Machining the Rear Covering

In Chapter 4, we attached a piece of pegboard to the inside of the case back with eight #4 sheet metal screws. We now need to temporarily remove that covering in order to machine holes to allow the control electronics cable and AC power cable to exit through the case rear. To do that, perform the following:

1. Mark the top of the rear covering with the letter "T" (to ensure you can reinstall it correctly).
2. Remove the eight screws holding the covering to the case rear.
3. Machine two notches in the rear covering as shown in *Figure 8-1*.
4. Place the rear covering and screws aside temporarily.

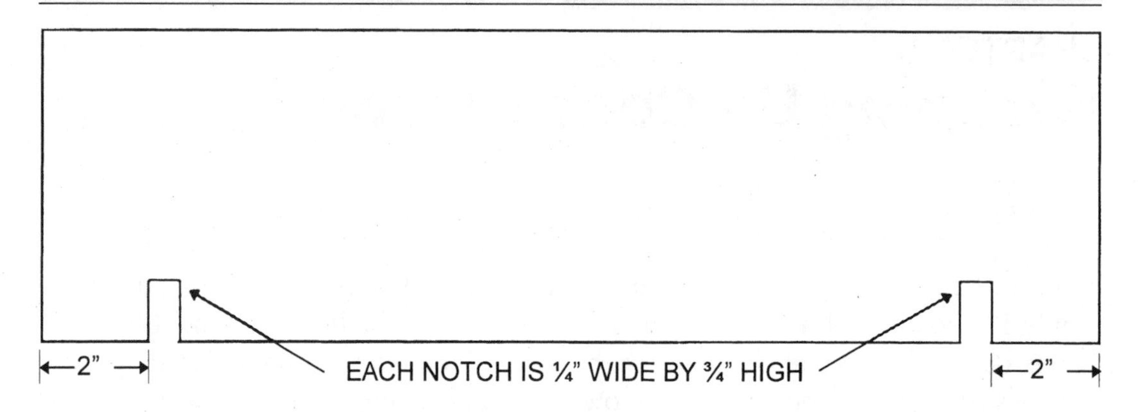

Figure 8-1. *Machine two notches in the rear covering as shown.*

Installing the Control Electronics

The control electronics will be attached to four screws mounted to the case base. A cable clip attached to one of the mounting screws will retain the cable. Refer to *Figure 8-2*:

1. Locate the control electronics so the PC board is at least one inch away from the case rear, and the cable can exit the notch in the case rear covering without bending sideways.
2. Mark the locations of the PC board mounting holes on the case bottom. At each marked location, drill a 1/8" diameter hole. Pass a #4-40 x 1" machine screw (#4) from the outside of the case, and secure the screw with a #4-40 nut (#5).
3. Install an additional #4-40 nut on each mounting screw and position about 1/4" above the surface of the case bottom.
4. Install the control electronics on the mounting screws and secure in place with additional #4-40 nuts.

Installing the Power Supply Electronics

The power supply electronics will be attached to two screws mounted to the case base. Refer to *Figure 8-2*:

1. Locate the power supply electronics so the PC board is at least one inch away from the case rear and about an inch away from the control electronics.

2. Mark the locations of the PC board mounting holes on the case bottom. At each marked location, drill a 1/8" diameter hole. Pass a #4-40 x 1" machine screw (#4) from the outside of the case, and secure the screw with a #4-40 nut (#5).
3. Install an additional #4-40 nut on each mounting screw and position about 1/4" above the surface of the case bottom.
4. Install the power supply electronics on the mounting screws and secure in place with additional #4-40 nuts.

Installing the Transformer

The transformer will be attached to two screws mounted to the case base. Refer to *Figure 8-2*:

1. Locate the transformer so it is at least one inch away from the case rear, and about an inch away from the power supply electronics.

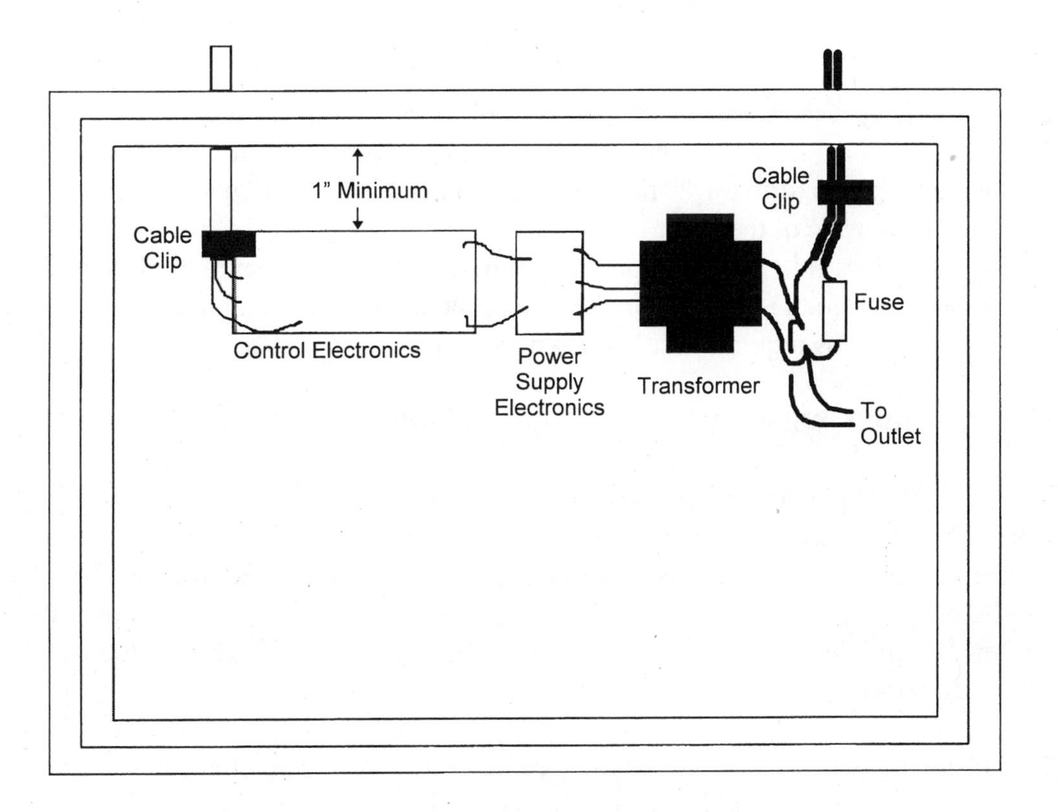

***Figure 8-2**. Installing the control electronics, power supply electronics, and transformer.*

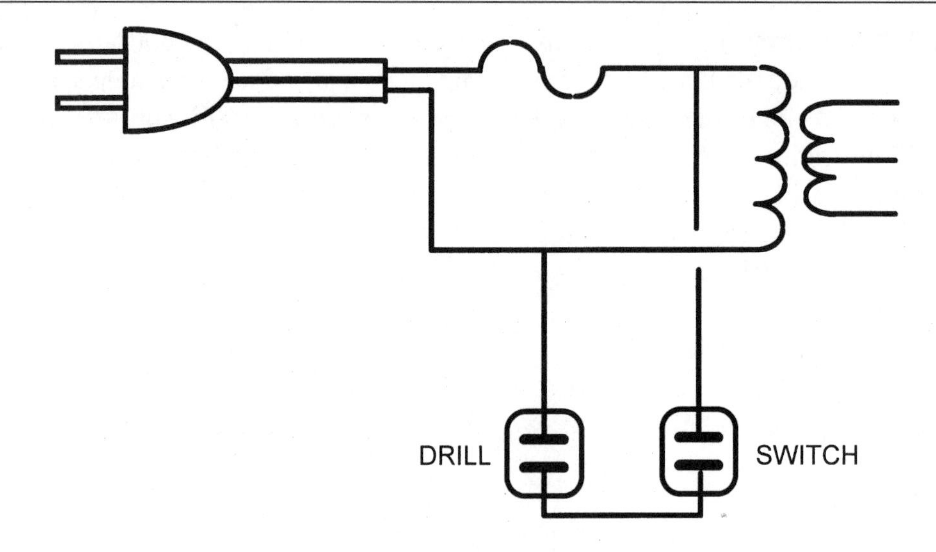

Figure 8-3. *Schematic for the final wiring.*

2. Temporarily disconnect the AC power cord from the transformer.
3. Mark the locations of the transformer mounting holes on the case bottom. At each marked location, drill a 1/8" diameter hole. Pass a #4-40 x 1" machine screw (#4) from the outside of the case.
4. Install the control electronics on the mounting screws and secure in place with #4-40 nuts (#5). Depending on the size of the transformer's mounting hole, flat washers may be required; use #4 or #6 flat washers as needed.

Modifying and Installing the AC Duplex Outlet

Next, the AC duplex outlet will be wired and installed in the rectangular hole in the base. The outlet has a brass colored set of screws on one side and a silver colored set of screws on the other. Either set of screws is electrically connected by a metal tab that can be broken off to electrically isolate the two screws. One of these tabs will be broken off, and the outlet will be wired to an 8" length of #18 line cord (sometimes called zip cord):

1. Locate the two brass screws on the side of the outlet. Using pliers, bend the tab between them back and forth until the tab breaks off.
2. Strip the two wires on one end of an 8" length of zip cord. Attach one wire to one of the brass screws. Attach the other wire to the other brass screw.

3. Pass the zip cord through the rectangular hole in the base so it exits on the bottom of the base. Install the outlet in the rectangular hole. Drill two 1/8" mounting holes in the outlet's two mounting ears. Secure the outlet to the base with two #4-40 x 3/4" machine screws (#3) and #4-40 nuts (#5).
4. Apply a duplex outlet cover plate to the duplex outlet with the screw provided with the cover plate.

Final Wiring

Perform final wiring as shown in *Figure 8-2*, and the schematic diagram of *Figure 8-3*:

1. Reinstall the case rear covering after passing the control electronics cable out of the case. That cable will exit through one of the notches in the rear covering.
2. From the outside of the case, pass the free end of the AC power cord through the remaining notch in the rear covering into the case.
3. As shown in *Figure 8-2* and *Figure 8-3*, connect one power cord lead to one lead of the in-line fuse. Secure with a wire nut.
4. Connect the other power cord lead to one of the transformer's primary coil leads. Also connect one of the leads from the duplex outlet to the other two leads. Secure all three wires with a wire nut.
5. Connect the remaining in-line fuse lead, transformer primary coil lead, and duplex outlet lead together and secure with a wire nut.

```
Current Speed Factor:  1

   PRESS      TO
   -----      ------------------------------
     2        Move Drill Stepper Clockwise
     4        Move Table Stepper Clockwise
     6        Move Table Stepper Counterclockwise
     8        Move Drill Stepper Counterclockwise
    Esc       End
   Enter      Change Speed Factor
```

***Screen 8-1**. The SPEEDSET screen.*

Constructing the Drill Power Switch

The drill power switch is simply a normally open push button switch mounted in a plastic enclosure, and connected to an AC line cord terminated in a standard AC plug. The switch is plugged into one of the duplex outlets and the drill is plugged into the other. When the switch is depressed, the AC circuit is completed and the drill is energized:

1. Obtain a small, all-plastic hobby box and a normally open push button switch. The switch should be rated for AC line voltage, typically 1 amp at 250 volts AC.
2. Machine an appropriately sized hole in the case top and then remove the case top. Mount the switch in the hole just machined using the hardware provided with the switch.
3. Obtain a standard household extension cord and clip off the outlet, leaving a length of zip cord terminated with an AC plug. Strip the insulation from the two wires on the free end. Form a knot about 6" from the free end.
4. Drill a 1/4" diameter hole in the side of the case. Pass the free end of the zip cord into the case. Form another knot on the inside of the case. Tighten the knots so the cord is secured against the case side.
5. Connect one wire to one of the switch's terminals (either terminal). Connect the remaining wire to the remaining switch terminal. Reinstall the case top and connect the plug to one of the duplex outlets in the *Auto-XY* base. Connect the power cord from the drill to the other duplex outlet.
6. Apply power to the *Auto-XY* power cord. Press and hold the drill power switch. The drill should energize. Release the switch and the drill should stop.

Installing the Base to the Case

Reinstall the base to the case, ensuring all wires are properly dressed and will not interfere with the movement of the case drawer. Secure the base to the case with several screws around the base's perimeter.

Adjusting the Setup File

In Chapter 5, we introduced the setup file (DRLSETUP.DAT) and modified it to include the address of your PC's port. We also introduced the SPEEDSET.EXE program which allows you to check for optimum speed setting. We'll now make one final adjustment to this file, refining the speed setting and ensuring the distance factor is correct.

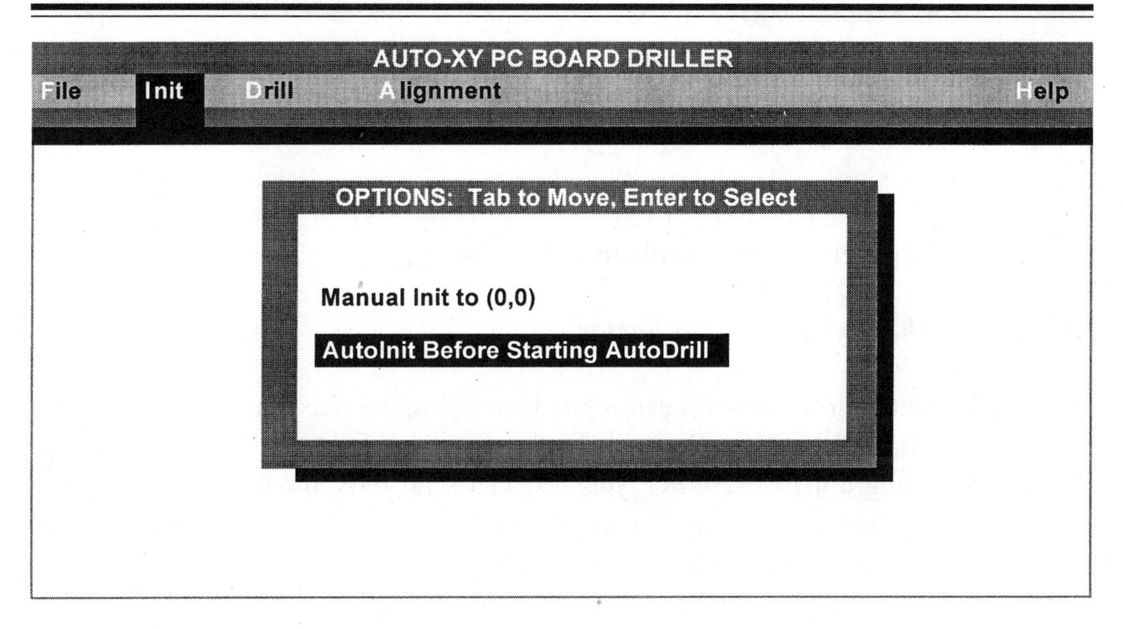

***Screen 8-2**. The Init screen.*

The Distance Factor

The third line in the setup file initially contains the number 864. To see where this number was derived from, we need to refer back to the characteristics of our stepper motor, and the threaded rod. The motor is specified as 7.5 degrees per step. Since a complete revolution traverses 360 degrees, the motor will make 48 (360 / 7.5) steps per revolution. The threaded rod specified is a 5/16" - 18, meaning there are 18 threads per inch. Therefore, to move one inch, the threaded rod has to make 18 complete revolutions. So,

$$48\frac{steps}{\cancel{revolution}} * 18\frac{\cancel{revolutions}}{inch} = 864\frac{steps}{inch}$$

The number 864 reflects how many steps need to be executed to move exactly one inch. The AUTOXY program uses that distance factor in all of its calculations. So, to ensure the accuracy of your hardware, do the following:

1. Determine the stepper motor's degrees per step. Call this D.
2. Determine the threaded rod's threads per inch. Call this T.

3. Solve the following equation to determine the distance factor F:

$$F = \frac{360}{D} * T$$

(Ex: For D = 7.5 and T = 18, F = (360/7.5) * 18 = 864)
Write down the results of your calculation here: **F =** ______________________

Determining the Optimum Speed Factor

The speed factor addresses two speed-related variables; the maximum speed at which the motor can operate, and the physical characteristics of the *Auto-XY* hardware. Unless you are using a 4.77 MHz PC, your PC can send movement commands much faster than the motors can accept. Therefore, a delay between movement commands is needed if the motor is to operate properly. In addition, the slight misalignments and other physical factors (like friction) further reduce the speed at which the motor can smoothly drive the drill and table.

The SPEEDSET program allows you to test various speed factors in real time and automatically revise the DRLSETUP.DAT file. Do the following to adjust the speed factor:

1. Power up your PC and connect the *Auto-XY* control cable (P1) to your PC's parallel port.
2. Power up the *Auto-XY*.
3. If your PC has a Windows based system, go to DOS.
4. Change directories to the AUTOXY subdirectory (type CD\AUTOXY and press Enter).
5. Type SPEEDSET and press Enter. You will see *Screen 8-1*.
6. Press Enter and change the speed factor to 1 (type 1 and press Enter).
7. Press 8 to move the drill towards the stepper motor. In all likelihood, the drill will not move and the drill stepper motor will hum. Press any key to stop "movement".
8. Multiply the current speed factor by 10 (ex: change 1 to 10, 10 to 100, etc.) and try again. NOTE: The maximum speed factor is 32767.
9. Note the speed factor at which the drill first begins to move. Divide that factor by 2, and try again. (For example, if the drill initially moves with a speed factor of 100, divide by 2 and try a new speed factor of 50.)
10. Repeat this process until the drill no longer moves smoothly. (For example, starting with 100, speed factors of 50 and 25 produce smooth movement, but a

speed factor of 13 causes the drill to bind.)

11. Try various speed factors between the current and previous factor. (For example, try values between 13 and 25.) Use the speed factor that allows both the drill and table to move smoothly without binding in both directions along their full movement distance.
12. Press <Esc>. You will see the question "Do you Want to Save the NEW Speed Data (Y/N)?…". Press Y to save the new speed data.

Modifying DRLSETUP.DAT

If you are using a standard parallel port and the 5/16"-18 threaded rod, your DRLSETUP.DAT file will look something like this:

```
888
15
864
Message
```

To change the file, do the following:

1. Type "COPY CON DRLSETUP.DAT" and press Enter.
2. Assuming the parallel port you want to use is at decimal address 888, type "888" and press Enter. If the parallel port you want to use is at another address (like 928 decimal), type its number instead of 888.

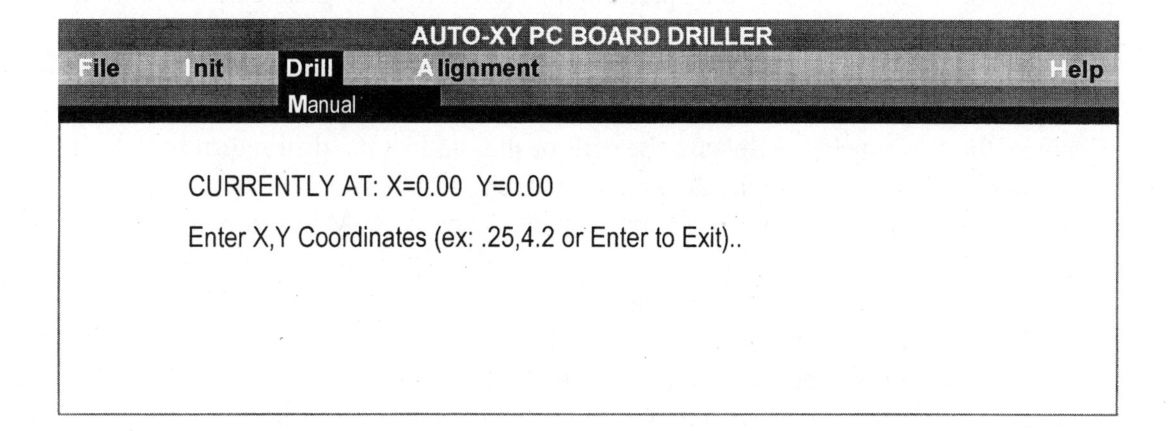

***Screen 8-3**. The Manual Drill screen.*

3. Type the appropriate speed factor (in this example, 15) and press Enter.
4. Type the value you wrote down for F above (in this example, 864) and press Enter.
5. Type the word Message and press Enter.
6. Press <Ctrl>Z, and then release both keys. The screen will show "^Z."
7. Press Enter. You will see the message "1 file(s) copied."

You can do this as often as you like. However, once you've specified the distance factor, the only time you'd have to change the setup file is if you planned to use *Auto-XY* on another PC with a different parallel port address. You can change the speed setting with SPEEDSET.EXE at any time, and it will automatically update the setup file.

Origin and X-Y Axes Marking

The last task we need to complete before being able to use *Auto-XY* effectively is to mark the origin, and the X and Y axes. The origin (0,0) allows us to accurately position a corner of a PC board to be drilled. The X and Y axes allow that PC board to be positioned parallel to movement of both the drill and table, ensuring accurate registration of the holes that will be drilled. To create the marking, do the following:

1. Remove the drill, install a drill bit in the drill and reinstall the drill in the drill caddy. (Make sure the drill bit is at least 1/4" above the table surface. If not, readjust the drill collar.)
2. Run the AUTOXY program (at the DOS prompt, type AUTOXY and press Enter).
3. Select the "Init" function by pressing <Alt> I. You will see *Screen 8-2*.
4. Select the Manual Init option by pressing the Tab key. Then press Enter to start the manual initialization to position 0,0.
5. You are now at the origin (location 0,0). To drill a hole at this location, press and hold the drill power switch while pushing down on the drill pusher until the drill bit enters the table surface. Release the drill pusher and let the drill return back to its resting position. Release the drill power switch.
6. Select the Drill function by pressing <Alt>D. Then press M to select manual drilling. You will see *Screen 8-3*.
7. Type 0,5 and press Enter to move the drill 5 inches from the origin. When it arrives, drill a hole at this location.
8. Type 7,0 and press Enter to move the drill back to the origin and move the table 7 inches from the origin. When it arrives, drill a hole at this location.
9. Press Enter to exit the manual drilling function.

10. Press <Alt>F to select the File functions. Press X to exit the program.
11. Draw a line on the table between position 0,0 and 0,5. This is the Y-Axis.
12. Draw a line on the table between position 0,0 and 7,0. This is the X-Axis.

Wrapping it Up

At last we've completed all of the construction and alignment. In the next chapter, we'll begin using the *Auto-XY* by fabricating a PC board replacement for the power supply board we previously constructed. It is small enough to be a good first project, but has all the elements needed to demonstrate the complete process of PC board fabrication using *Auto-XY*.

CHAPTER 9
Using *Auto-XY* to Fabricate a PC Board

Introduction

In this chapter, we'll go through each of the steps in making a PC board using three different approaches:

1. MANUAL METHOD. In this approach, we'll design a PC board on paper and identify the X and Y positions of each hole. Then we'll create a .DRL (drill location) file that *Auto-XY* can use to drill the holes. We'll use those drilled hole locations to manually draw the PC pattern with a resist pen. Then we'll etch the board.
2. TRANSFER METHOD. In this approach, we'll use a toner transfer system. (This process uses special paper on which the reversed image of a PC board is impressed, either with a laser printer or photocopier.) We'll describe the differences between this and the manual method.
3. CAD METHOD. In this approach, we'll discuss the Excellon drill file created by most CAD packages, and explain how to convert them to a .DRL file format.

As we step through these methods, we'll learn about the .DRL file format, and proper positioning of the PC blank on the *Auto-XY* table.

Positioning a PC Board Blank on the Table

Prior to drilling, you will position the PC blank on the table and secure it. You want one corner of the blank to be right at the (0,0) position. You also want one edge to be parallel to the drill movement (Y-axis), and the other edge to be parallel to the table movement (X-axis). In this way, the resultant hole pattern will be correctly located on the blank.

If you place the blank directly on the table, you'll be drilling through the blank and into the plywood. This is undesirable for two reasons. The table's plywood has a grain that can cause the drill bit to follow that grain, twisting the brittle portion of the bit and causing it to break. Also, after repeated use, there would be many holes in the table. If

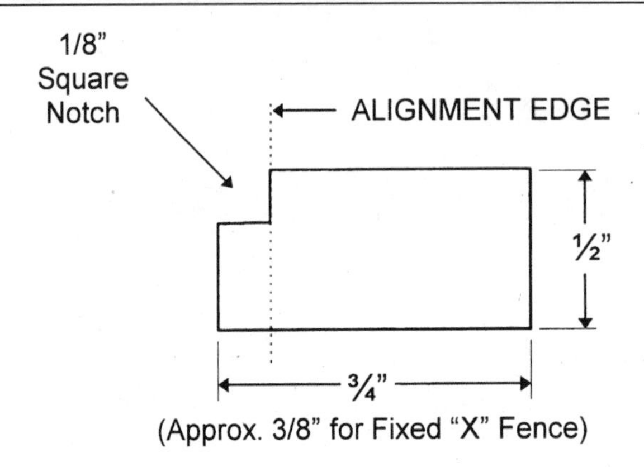

Figure 9-1*. Cut a 1/8" square notch on one edge of the blank.*

the bit enters a previously drilled hole off-center, the bit can also twist and break. Placing sacrificial cardboard under the blank solves this problem, but creates another. The cardboard flexes, and can cause the blank to move down at an angle against the cardboard. This can also cause the bit to twist and break.

A better solution is to create fixed and movable fences on which to mount the blank. To do this, create various lengths of fence from miscellaneous wood stock. To create the fence stock, machine wood stock to an overall size of 1/2" thick by 3/4" wide. Then make a saw cut on one edge such that the resulting notch is about 1/8" square (as shown in *Figure 9-1*). Cut the fence stock to the following lengths:

1. One piece 8 3/4" long (fixed "X" fence).
2. One piece 6" long (fixed "Y" fence).
3. Two each pieces 1", 2", 3", 4", and 5" long (movable fences).

If you look at the position for the "X" fence, you'll notice that its 3/4" width will interfere with the table threaded rod driver. So, cut this fence down to a width of about 3/8". Test fit it on the table and make any further width adjustments to ensure it does not interfere with the rod.

Now, referring to *Figure 9-2*, align the "X" fence on the table, using the alignment edge shown in *Figure 9-1* (not the actual edge of the fence). Make sure the alignment edge is directly over the line you drew during calibration. Note that the end of the "X" fence overruns the Y = 0 line by 3/4" (the width of the fence stock). Clamp in place and drill two mounting holes through the fence and the table. Remove the fence and enlarge the holes in the table to allow some adjustment. Secure the fence to the table with machine screws, nuts, and washers. Repeat this process for the "Y" fence, butting the end of this fence to the previously installed "X" fence.

To test out the fences, use the Alignment function of the AUTOXY program. Move the drill and table to their zero positions. With a bit in the drill, power the drill and ensure the bit comes down at the corner defined by the intersection of the "X" and "Y" fence alignment edges (the 0,0 position). Then move the drill near the end of the "Y" fence and ensure that the drill comes down on the alignment edge. Finally, move the drill back to the zero position and move the table near the end of the "X" fence. Ensure the drill comes down on the alignment edge. Make any adjustments to the fences as necessary.

To use the movable fences, position the PC blank on the fixed fences. Then, select appropriately sized movable fences and use them to support the two unsupported edges of the blank. Hold them in place with masking tape. Finally, secure the four corners of the blank with masking tape.

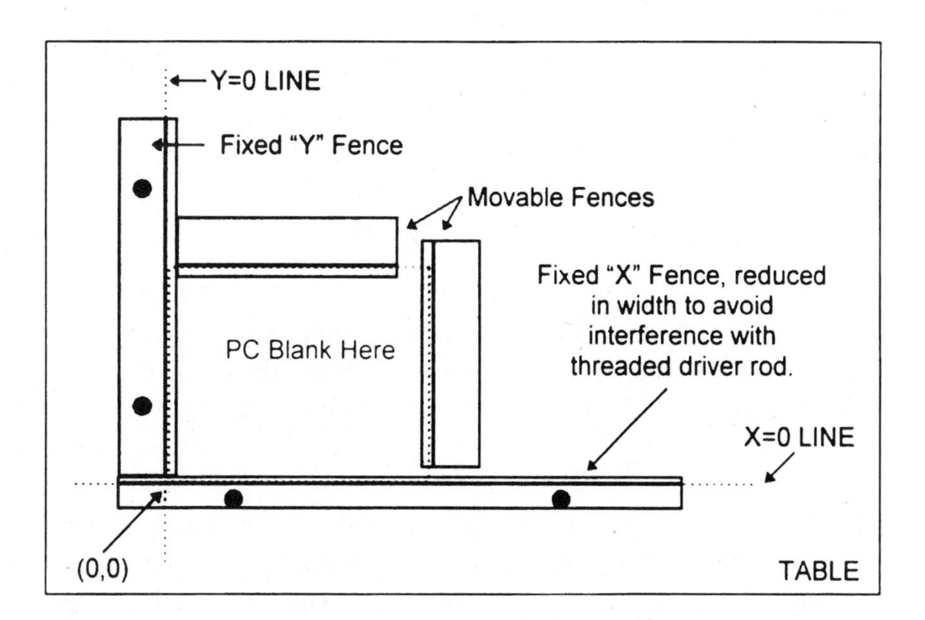

Figure 9-2*. Align the "X" fence on the table as shown.*

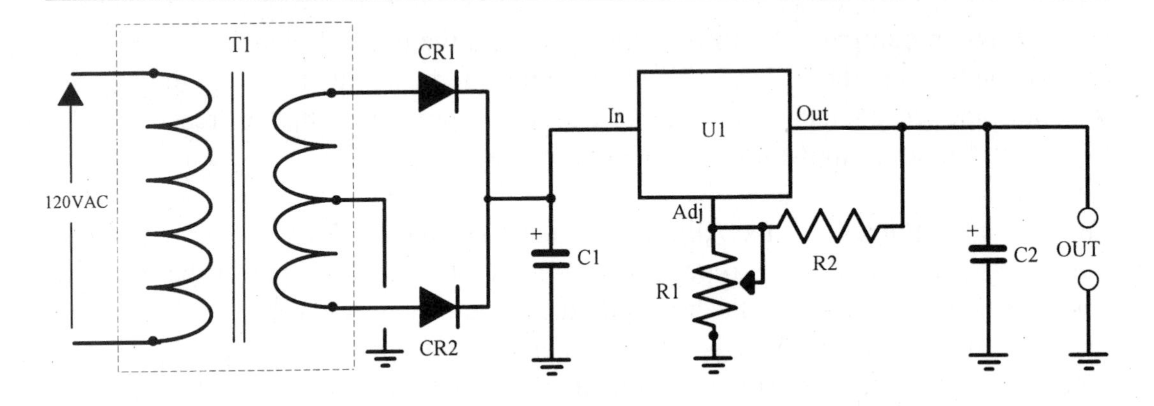

Figure 9-3. *The schematic of the Auto-XY power supply.*

The Manual Method

This process is well suited for those one-of-a-kind designs that you want to get onto a PC board. There are a few items you'll need if you plan to use this method. They are:

1. Circle template.
2. Resist pen.
3. PC board blank stock.
4. 0.1" graph paper (or a PC drawing application).

Step 1

The first step is to design the actual PC pattern on scratch paper. *Figure 9-3* shows the schematic of the *Auto-XY* power supply. *Figure 9-4* shows the initial design on scratch paper. The darker lines are the PC pattern. The lighter lines show the component placement.

At this point, you should check out the pattern to ensure it accurately reflects the circuit. Then you need to use the pattern to create an accurately sized representation of the PC board. This can be done on 0.1" grid paper or by using a drawing program. The 0.1" grid paper is well suited for this task, since most electronic component mounting is in increments of 0.1". The following are typical examples of standard component spacing:

1. Resistor: 0.4" (1/4 watt), or 0.5" (1/2 watt).
2. Capacitor: 0.3" (disc), 0.1" to 0.3" (radial electrolytic).
3. Transistor: 0.1".
4. Integrated circuit: 0.1" (0.3" between rows).

Step 2

Figure 9-5 shows the pattern redone on a 0.1" grid. While this could have been done on grid paper, this particular layout was created in Microsoft Word using its drawing tools. A comparison of *Figures 9-4* and *9-5* show the same interconnections have been maintained, but the spacing and line routing have been adjusted to conform to the component spacings identified above.

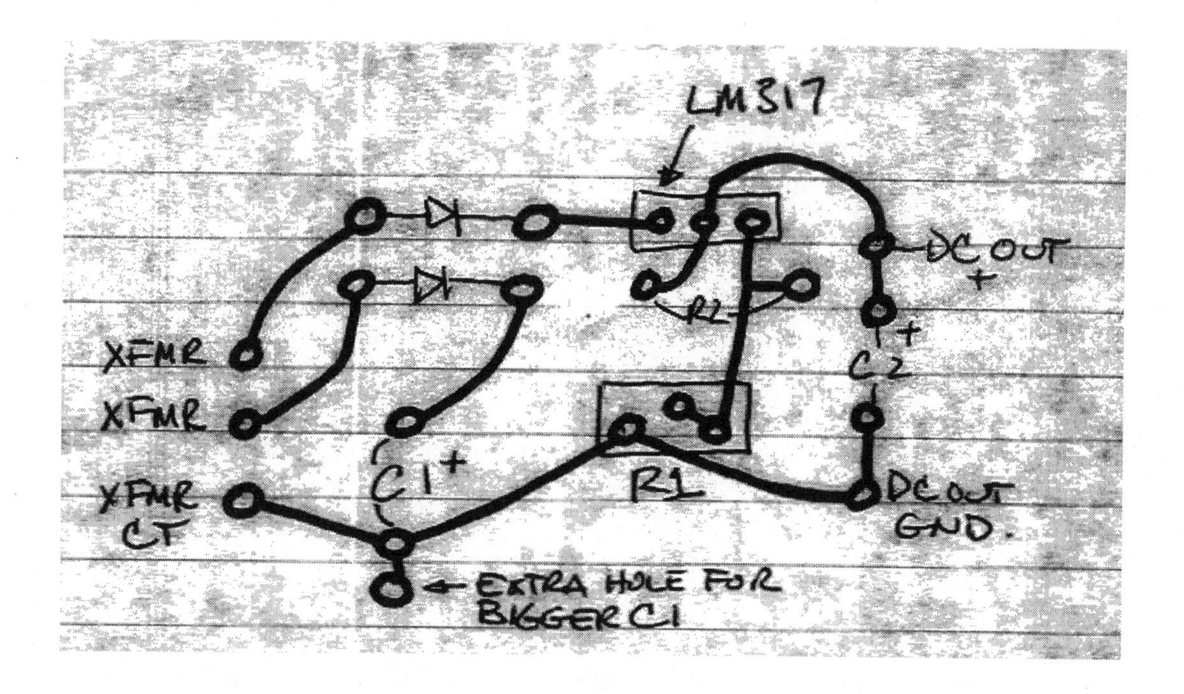

***Figure 9-4**. The initial PC pattern design sketch.*

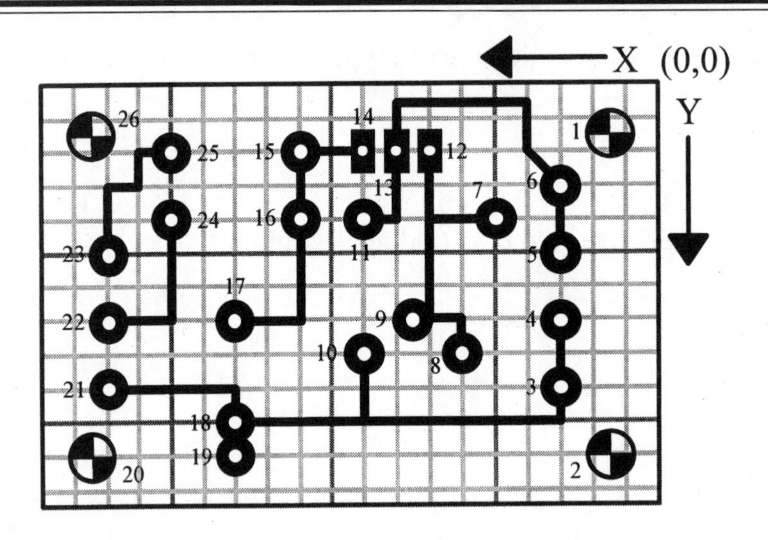

Figure 9-5. *The PC pattern redone on a 0.1" grid.*

Step 3

In this step, we'll identify an X and Y location for each of the holes. Before we begin, we must arbitrarily identify the origin. *Figure 9-5* shows that origin specified as the upper right hand corner, although it could just as well have been any of the four corners.

Also identified are the X-axis and Y-axis. This is important, since the hole coordinates are (X,Y). Once the origin is set, the X-axis must go horizontally from the origin along the board. The Y-axis must go vertically from the origin along the board.

The last identifications in *Figure 9-5* are numbering of each hole. Although not absolutely necessary, hole numbering helps avoid confusion and error as each hole location is identified in a .DRL listing.

Step 4

In this step, we'll identify the X and Y location of each of the holes. Each hole's X and Y location is read off the layout using the grid as the reference. As an example, *Figure 9-6* shows that hole 5 is located 0.3" in the X direction from the (0,0) point, and 0.5" in the Y direction from the (0,0) point. As a result, hole 5's location would be indicated as 0.3, 0.5 in the overall listing of hole locations. In practice, the leading zeros, and any space between the comma are not really necessary. Therefore, the listing would have the location listed as .3,.5. The resultant tabulation of all holes on the PC board is shown in *Figure 9-7*.

Step 5

In this step, we'll convert the tabular listing of *Figure 9-7* to the .DRL file format. That format is:

First Line	ID OR TITLE LINE	
Second Line	X,Y	:overall dimensions
Third Line	X,Y	:location of hole 1
Remaining Lines	X,Y	:location of remaining holes (one per line)

The first and second lines are informational, and will be displayed when you select AUTOXY's auto drilling function (after first opening the file with its FILE and OPEN functions).

The remaining lines contain the X and Y locations of each hole, with the Y location being separated from the X location by a comma. This can be followed by a colon (or any other non-number character) and descriptive text. That text will not be used by AUTOXY, but it does serve to document your .DRL file, should you need to change it at a later date. The separator after the X,Y values and before the descriptive text also applies to the second line containing the PC board overall dimensions.

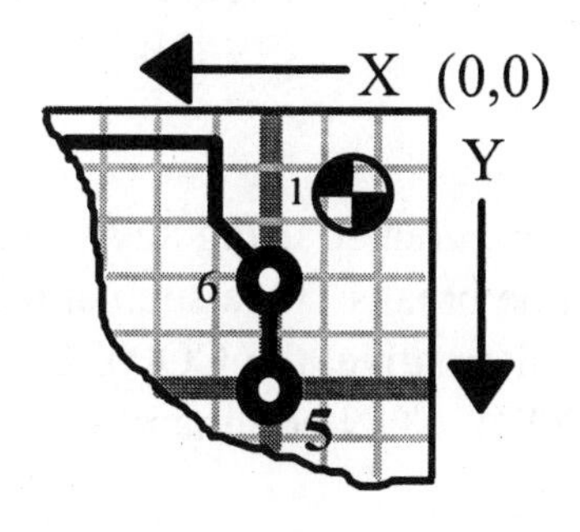

***Figure 9-6**. The location of hole 5.*

Hole	Coordinates
Hole #1	.15,.15
Hole #2	.15,1.1
Hole #3	.3, .9
Hole #4	.3,.7
Hole #5	.3,.5
Hole #6	.3,.3
Hole #7	.5,.4
Hole #8	.6,.8
Hole #9	.75,.7
Hole #10	.9,.8
Hole #11	.9,.4
Hole #12	.7,.2
Hole #13	.8,.2
Hole #14	.9,.2
Hole #15	1.1,.2
Hole #16	1.1,.4
Hole #17	1.3,.7
Hole #18	1.3,1
Hole #19	1.3,1.1
Hole #20	1.75,1.1
Hole #21	1.7,.9
Hole #22	1.7,.7
Hole #23	1.7,.5
Hole #24	1.5,.4
Hole #25	1.5,.2
Hole #26	1.75,.15

Figure 9-7. *Tabulation of all holes on the PC board.*

```
Auto-XY Power Supply PC Board
1.9,1.25     :Overall Board Dimensions
.15,.15      :Hole #1, Mounting
.15,1.1      :Hole #2, Mounting
.3, .9       :Hole #3, DC Out - Gnd
.3,.7        :Hole #4, C2-
.3,.5        :Hole #5, C2+
.3,.3        :Hole #6, DC Out - +
.5,.4        :Hole #7, R2
.6,.8        :Hole #8, R1
.75,.7       :Hole #9, R1
.9,.8        :Hole #10, R1
.9,.4        :Hole #11, R2
.7,.2        :Hole #12, LM317 - Adj
.8,.2        :Hole #13, LM317 - Out
.9,.2        :Hole #14, LM317 - In
1.1,.2       :Hole #15, CR1 Cathode
1.1,.4       :Hole #16, CR2 Cathode
1.3,.7       :Hole #17, C1+
1.3,1        :Hole #18, C1-
1.3,1.1      :Hole #19, Extra C1 Hole For Larger Cap
1.75,1.1     :Hole #20, Mounting
1.7,.9       :Hole #21, Xfmr Center Tap
1.7,.7       :Hole #22, Xfmr Secondary
1.7,.5       :Hole #23, Xfmr Secondary
1.5,.4       :Hole #24, CR2 Anode
1.5,.2       :Hole #25, CR1 Anode
1.75,.15     :Hole #26, Mounting
```

Figure 9-8. *The completed POWERSUP.DRL file.*

Since the file is plain text, it can be created using any word processor that can save the document as ASCII text with line breaks. One such application available to all Windows users is *Notepad*. As an alternative, the COPY CON command in DOS can be used to create the file. (The COPY CON command was used in Chapter 8 to modify the DRLSETUP.DAT file.)

When creating the file, be sure to avoid blank lines at the file's end, as these will be interpreted as additional holes. Save the file as plain text with line breaks, and specify a file extension. AUTOXY will allow you to specify any file name (including any file name extension) you wish. To be consistent, it is recommended that you name these files using the .DRL extension.

The completed file, POWERSUP.DRL, is shown in *Figure 9-8*. The file was created using Windows *Notepad* and saved as a textfile named POWERSUP. Since Notepad (and some other word processors) automatically add a .TXT extension during file save, the file was renamed after saving to POWERSUP.DRL. Note that, in addition to the hole numbers, additional descriptive text has been added to indicate which component lead populates each hole.

Step 6

Now it's time to drill the holes in the PC blank. Cut a 1.9" x 1.25" piece of single sided PC board material from a larger piece. Position it on the table with the 1.9" dimension along the X-axis, and the 1.25" dimension along the Y-axis. Tape the blank down at all four corners with masking tape.

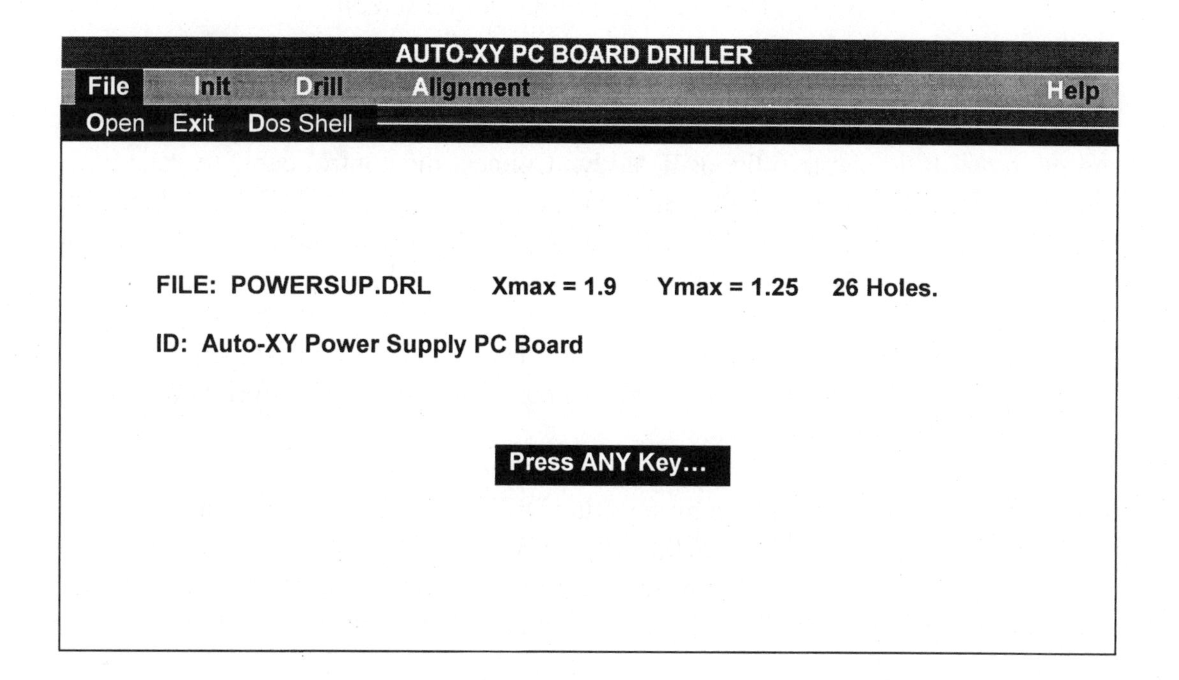

Screen 9-1*. The initial POWERSUP.DRL screen.*

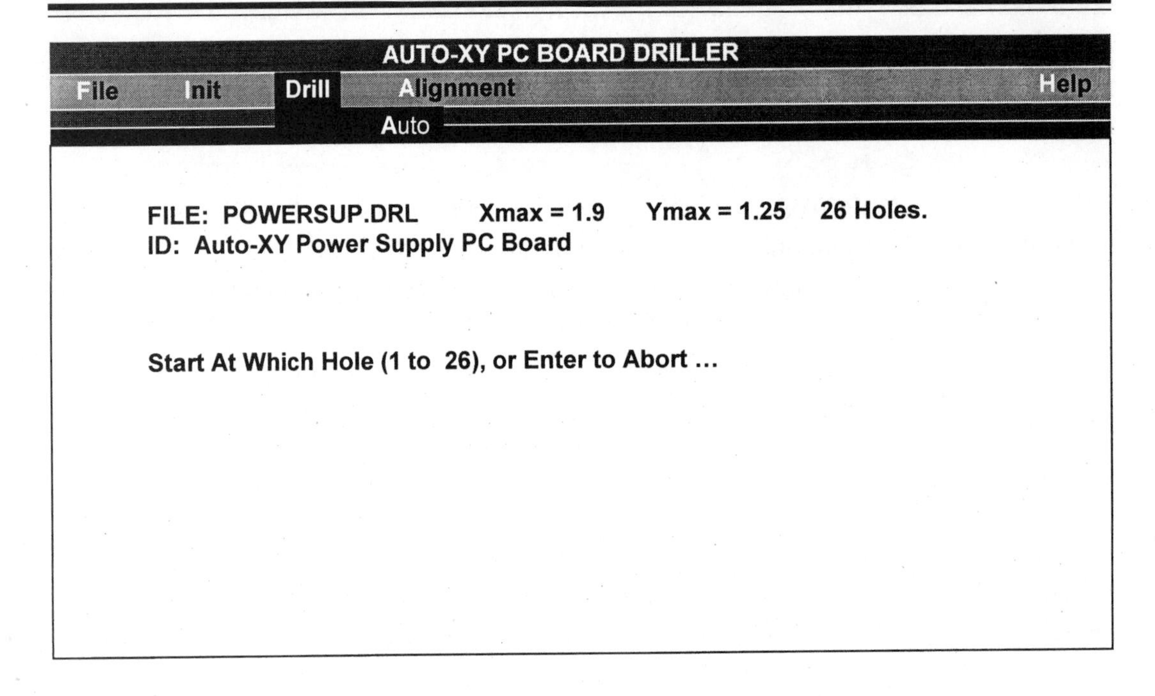

Screen 9-2. *The drill initialization screen.*

Remove the drill from the drill caddy, chuck an appropriately sized bit into the drill, and place the drill back into the drill caddy. Connect the control cable to your PC's parallel port, and power up your PC. Bring the PC to DOS in the AUTOXY subdirectory. Power up *Auto-XY* and run the AUTOXY program. At the start screen, press any key as indicated under the startup message.

Select the POWERSUP.DRL file by selecting File (<Alt>F) and then Open (O). Type in the name (POWERSUP.DRL) and press Enter. You will see the information shown in *Screen 9-1*. Press any key.

Now, select the auto drill function by selecting Drill (<Alt>D) and then Auto (A). You will see the message "Initializing to 0,0. Please Wait..." while the table and drill travel towards their zero position. When initialization has been completed, you'll see the information in *Screen 9-2*. Press 1 and then Enter to begin drilling at hole #1.

As the drill moves to hole #1, you'll see the message "Moving to Hole 1 (0.15,0.15)". When hole #1 is reached, you'll see another message appear: "ARRIVED. Press Any

Key for Next Hole..." In addition, at the bottom of the screen in red is the message "Esc To Abort." At any time if you want to abort, simply press <Esc>.

To drill the hole, press the drill's power push button switch to start the drill turning. Then press down with your finger on the drill pusher (specifically, in the 1/4" hole in the pusher). Drill the hole and then release the pusher. Finally, release the drill's power push button switch. Repeat this process for all 26 holes. You'll notice as the drill approaches hole 3, it will move away from the stepper and at the very end of travel will quickly reverse direction for a short distance. You'll see the same "lash back" movement of the table as it approaches hole 12. This is called lash back correction, and corrects for any minor pivoting of the drill or table about its driver nut. Lash back correction is explained in more detail as part of the *Auto-XY* Program discussion in Chapter 10.

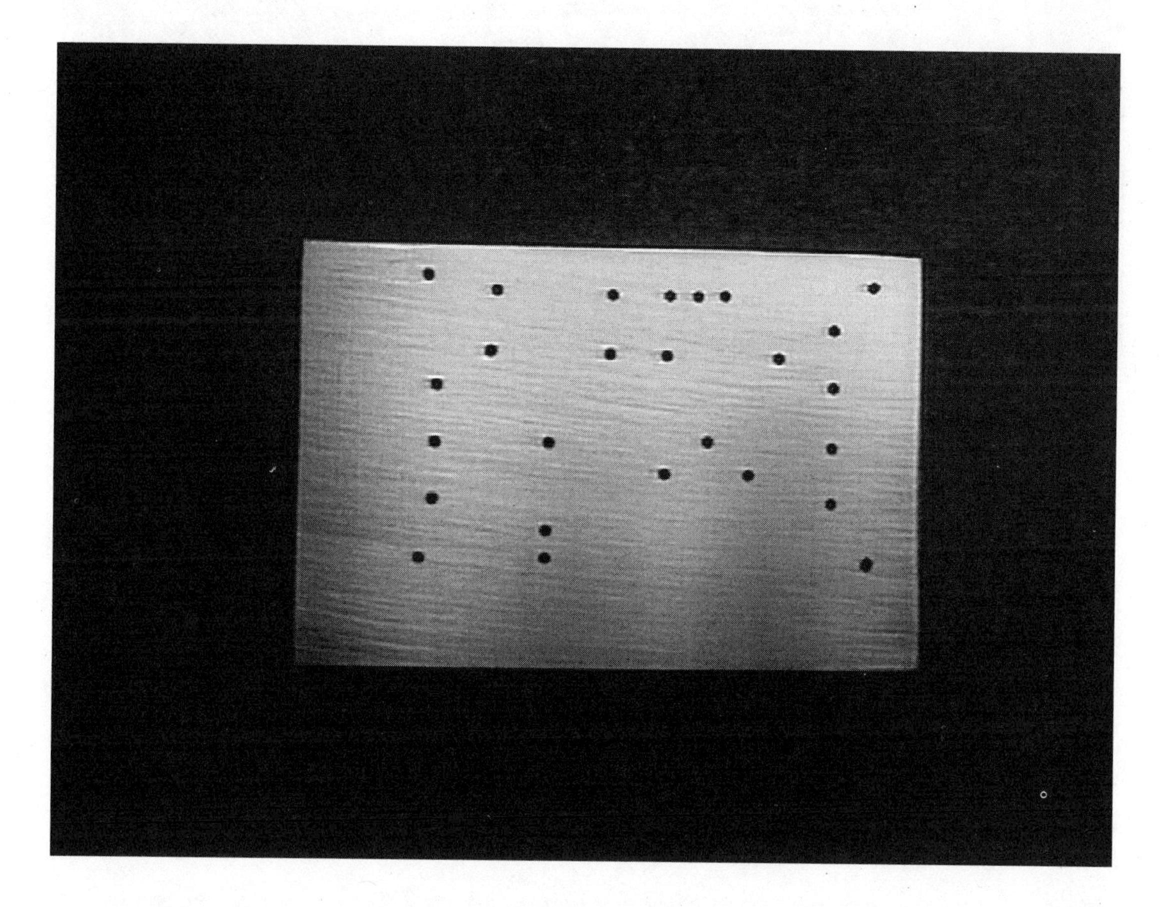

***Photo 9-1**. The drilled blank.*

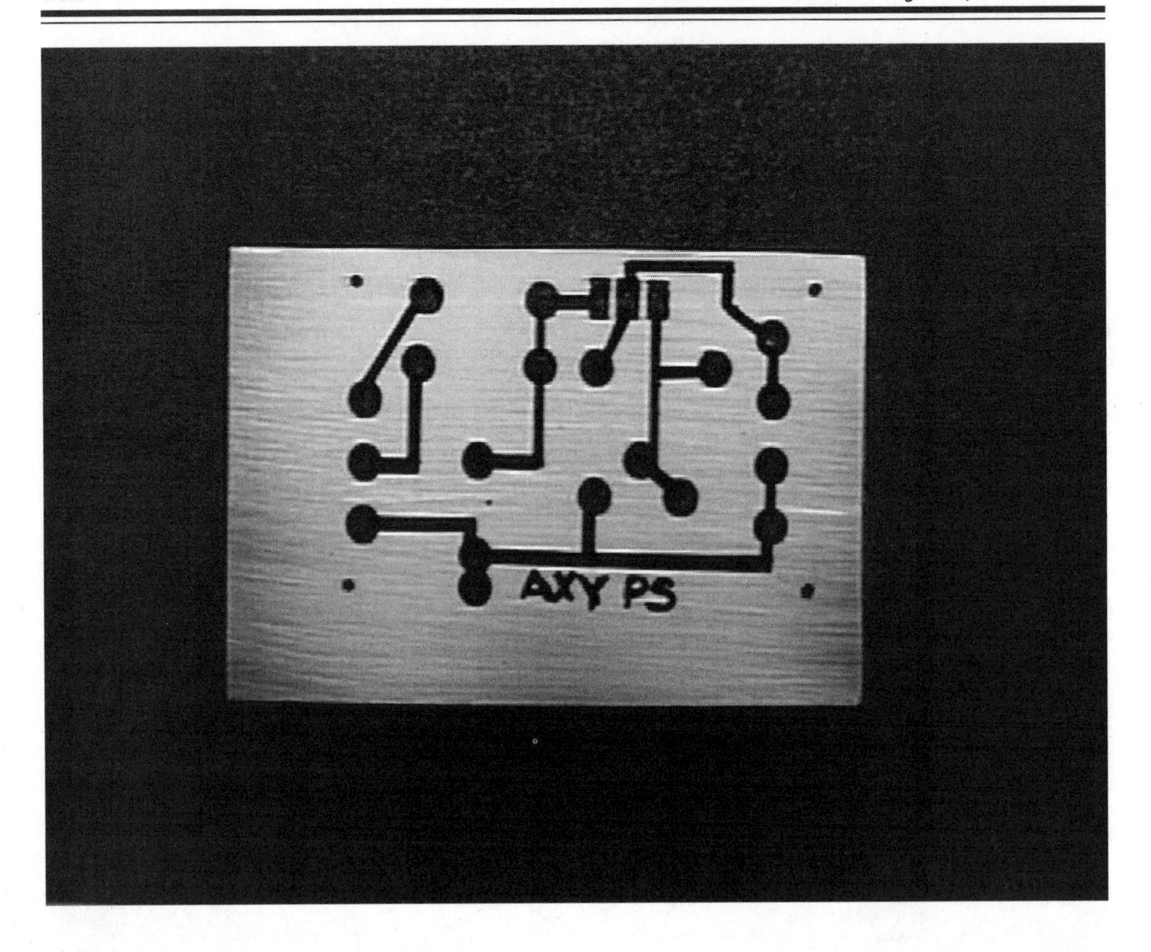

***Photo 9-2**. The PC board is ready for etching.*

When done, remove the blank from the table. To end the program, select File <Alt>F and then Exit (X). *Photo 9-1* shows the drilled blank.

Step 7

The drilled holes in the blank will now be used as location points for the PC pattern that will now be drawn on the blank with a resist pen and a circle template.

Using an abrasive pad, non-soap filled scouring pad, or 220 grit sandpaper, scour the copper to remove any burrs and oxidation. *Do not touch the copper surface after this scouring, as the oils from you hand will inhibit the resist pen ink from adhering to the copper.* Secure the blank to a firm surface by applying a small piece of masking tape to each corner.

Locate your circle template and resist pen. Center an appropriately sized hole in the template over a hole in the blank, flexing the template so it does not contact the blank over the hole (the resist ink will spread under the template if it is in contact with the blank).

Position the resist pen's tip in the hole and draw a circle around the hole in the blank, filling in the area to the perimeter of the drilled hole. The resist should form a continuous pattern around the hole. (There is no need to draw a resist circle around the four mounting holes #1, #2, #20, and #26.) Form a resist pattern around the remaining 21 holes.

Referring to the PC layout, draw in all circuit traces. Be sure that the trace lines are thick enough to withstand the etching process, and the resulting lines are dense black. If necessary, go over the lines after they have dried to create a dense black line.

Check your work to ensure you have drawn in the complete layout, and have not misconnected any circuit traces. *Photo 9-2* shows the board ready for etching.

Etching a PC Board

Pour enough etchant solution into a *plastic* tray to cover the blank at least 1/8". Place the blank in the tray, copper side up, and push it down so the etchant covers it. NOTE: Heating the etchant prior to placing the blank in it will speed the etching process. However, *extreme caution should be observed, since the etchant is a caustic solution and could cause property damage or personal injury!!*

Agitate the solution with a wood or other nonmetallic probe. Periodically angle the blank out of the etchant solution to observe the etching progress. When all excess copper has been etched away, transfer the plastic tray to a sink with running water.

Remove the blank from the tray and wash in cold running water for at least one minute. Dispose of the etchant solution, and thoroughly wash the plastic tray and probe. Using a soap filled scouring pad, remove the resist ink from the blank. Pat the blank dry with a paper towel. The PC board is ready for application of components. *Photo 9-3* shows the completed PC board.

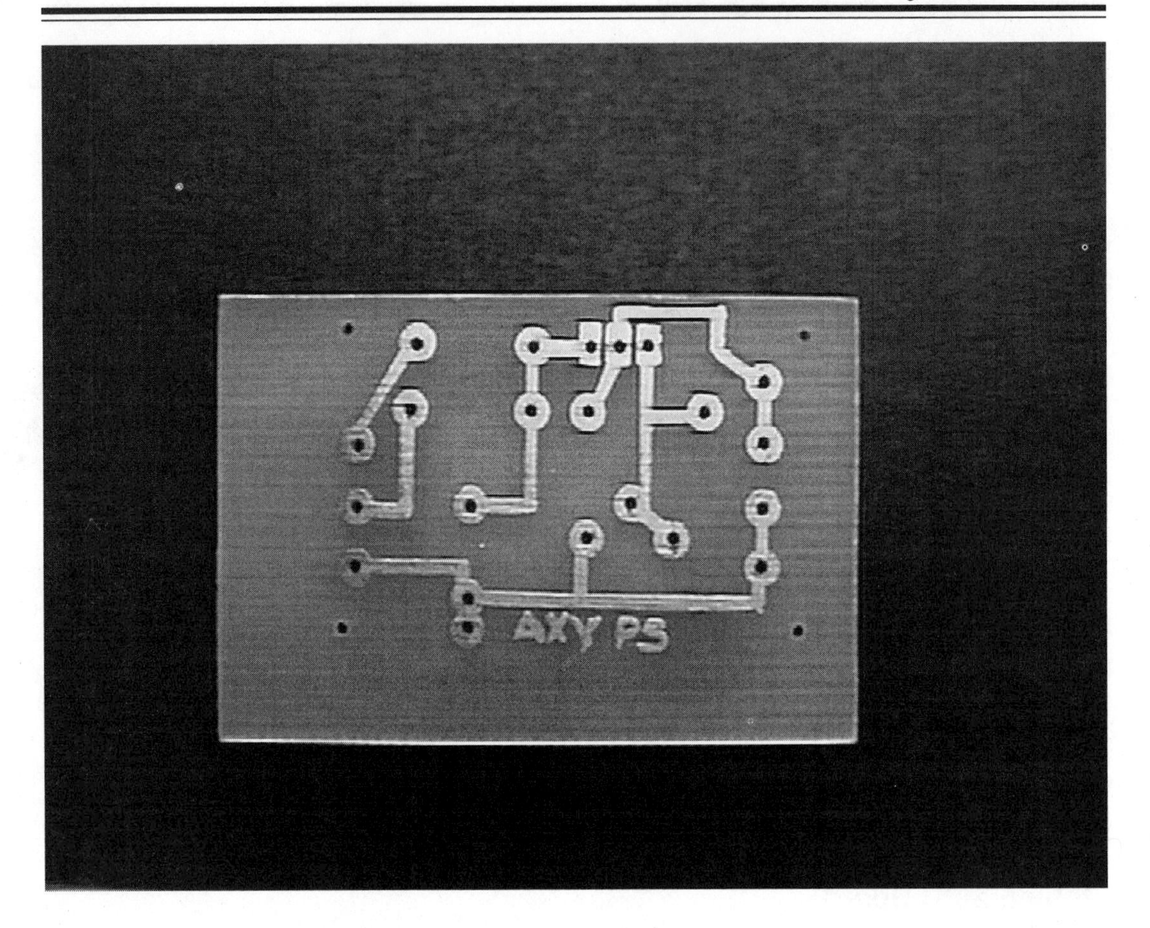

Photo 9-3. *The completed PC board.*

The Toner Transfer Method

This method is useful if you plan to fabricate a number of identical boards. Toner transfer material is available from several sources, one being DynaArt Designs (805-943-4746). For this process, you'll need:

1. Toner transfer material.
2. Laser printer or photocopier.
3. Heat source (like an iron).

Similarities with the Manual Method

Steps 1 (preparing the layout) through 6 (drilling the PC blank) are identical for both the manual and toner transfer methods. However, for the toner transfer method, it is best to create the PC pattern using a computer-based method.

Preparing the Pattern for Transfer

Following the manufacturer's directions, use your computer application to "flip" the pattern horizontally (resulting in a reversed pattern). Print out that pattern on the toner transfer material using a laser printer.

Transferring the Pattern

Cut out the pattern and position it over the board, aligning the holes in the blank with the circles on the pattern. Follow the manufacturer's instructions for transferring the pattern to the blank.

Etching the Board

Etch the board as previously described.

The CAD Method

The CAD method uses CAD software to create the PC pattern. Then, the CAD software can make a flipped printout on toner transfer material using a laser printer.

Most CAD packages will create a drill list called an "Excellon" file. Rather than manually transfer the hole locations from the CAD package to a .DRL file, a conversion program can be created to convert the information in the excellon file to the .DRL format.

Program Listing 9-1 shows such a conversion program written by an *Auto-XY* user, Mark Bianchi of Yukon, Oklahoma:

```
'*************** EXCELLON TO .DRL FILE CONVERTER ***************
'    INTRO SCREEN
CLS : COLOR 15, 0
LOCATE 1, 1: PRINT STRING$(80, 196): COLOR 14
```

```
LOCATE 2, 15: PRINT "EXCELLON TO X/Y COORDINATE DRILL FILE
CONVERTER "
LOCATE 3, 30: PRINT "By Mark Bianchi": COLOR 10
LOCATE 5, 16: PRINT "To Convert Excellon Files To .DRL Format
For"
LOCATE 6, 18: PRINT "Auto-XY Designed By James J. Barbarello"
LOCATE 7, 1: COLOR 15, 0: PRINT STRING$(80, 196)
'************************* CODE *************************

LOCATE 9, 1: INPUT "Enter Input File (.exl) "; inputfile$
LOCATE 11, 1: INPUT "Enter Output file (.drl)"; outputfile$
COLOR 7, 0
LOCATE 13, 5: PRINT "Enter Descriptive Text (50 Char Max)."
LOCATE 14, 5: INPUT descriptivetext$
LOCATE 16, 5: INPUT "Enter board size (i.e. 5.25,3.50)"; maxx,
maxy
OPEN inputfile$ FOR INPUT AS #1
OPEN outputfile$ FOR OUTPUT AS #2
OPEN fo$ FOR OUTPUT AS #2

WRITE #2, descriptivetext$
WRITE #2, maxx, maxy

ret:
IF EOF(1) = -1 THEN GOTO over
INPUT #1, a$
c$ = MID$(a$, 1, 1)
IF c$ = "/" OR c$ = "" OR c$ = "T" OR c$ = "M" THEN GOTO ret
x$ = MID$(a$, 2, 2) + "." + MID$(a$, 4, 2)
y$ = MID$(a$, 9, 2) + "." + MID$(a$, 10, 2)
PRINT VAL(x$); ","; VAL(y$)
WRITE #2, VAL(x$), VAL(y$)
GOTO ret

over:
CLS : LOCATE 10, 36: PRINT "DONE!"
END
```

The program accepts input in the form of the existing excellon file name, the target .DRL file name, the first line of descriptive text to appear in the .DRL file, and the overall board size. Then the program extracts the X,Y locations of each hole from the excellon file, formats that information, and passes it to the .DRL file. The program contains no error trapping, so be sure to include the full excellon and .DRL file names (including the file extensions and any paths). Let's take a look at the conversion process:

```
1  ret:
2  IF EOF(1) = -1 THEN GOTO over
3  INPUT #1, a$
4  c$ = MID$(a$, 1, 1)
5  IF c$ = "/" OR c$ = "" OR c$ = "T" OR c$ = "M" THEN GOTO ret
6  x$ = MID$(a$, 2, 2) + "." + MID$(a$, 4, 2)
7  y$ = MID$(a$, 9, 2) + "." + MID$(a$, 10, 2)
8  PRINT VAL(x$); ","; VAL(y$)
9  WRITE #2, VAL(x$), VAL(y$)
10 GOTO ret
```

This is the loop from the conversion program, beginning with line 1 and ending with line 10. Line 2 checks for the end of file and exits the loop when found. Line 3 gets a line from the excellon file. Line 4 takes the first character of the line and puts it in the variable c$. Line 5 checks to see if that character indicates the line is not an X,Y location. If so, it skips over that line.

The excellon file contains lines with the X and Y locations in the form "X0123 Y0456." Line 6 takes the first two digits after the "X" indication and identifies this as the whole number part of the X location. It then takes the next two digits and identifies this as the decimal part of the X location, placing a decimal at the appropriate point. So, for the form shown, line 6 would convert this to 1.23. Line 7 does the same for the y value, resulting in a conversion like 4.56. Line 8 takes these two values and writes them to the .DRL file in the proper format (ex: 1.23,4.56).

Wrapping it Up

In this chapter, we've added supporting fences to the table to provide precise positioning of the PC blank, and to minimize the possibility of drill bit damage. We've gone through the various methods of creating a PC pattern and using that pattern to fabricate a PC board. In the chapters that follow, we'll take a more in-depth look at the software (including the AUTOXY functions, and the actual program code). We'll finish off with some thoughts on how to modify the *Auto-XY* to add a Z-axis control, making drilling totally automatic.

CHAPTER 10
The Software

Introduction

It's now time to take a closer look at the AUTOXY program, and how it works. We'll begin by describing each of its functions and sub-functions. Then we'll take a look at two other programs, FINDPORT and SPEEDSET to see how to find an available parallel port, and provide the proper instructions to the control electronics.

All three software programs (FINDPORT.EXE, SPEEDSET.EXE, and AUTOXY.EXE) are executable files originally written in Microsoft QuickBasic. The source code files (*.BAS) are included on the companion disk for reference only, since they aren't needed for operation of the *Auto-XY*. The source code listing for each of these programs is also included here (*Program 10-1* through *10-3*) for your convenience.

Two additional files (DRLSETUP.DAT and DRLHELP.DAT) are required for proper operation of AUTOXY, and are also included on the disk. We've previously discussed DRLSETUP.DAT, its contents, and how to modify it. DRLHELP.DAT is a plain text file used by AUTOXY to display non-context sensitive help information when the Help function is selected.

Loading the Software

If you've already gone through the previous chapters, you'll have already loaded and used the software on the companion disk. If you've chosen to jump to this chapter to find out about the software, here's a procedure for loading the companion disk contents on your PC:

1. Insert the companion disk into floppy disk drive A: (or B:).
2. Go to your root directory (you can usually do this by typing C:\ (or D:\) and pressing Enter). If you're using a PC with Microsoft Windows, go to DOS (by exiting Windows 3.1, or selecting Start, Programs, and MS DOS Prompt in Windows 95) and then go to your root directory.
3. Make a new directory by typing MD AUTOXY and pressing Enter.

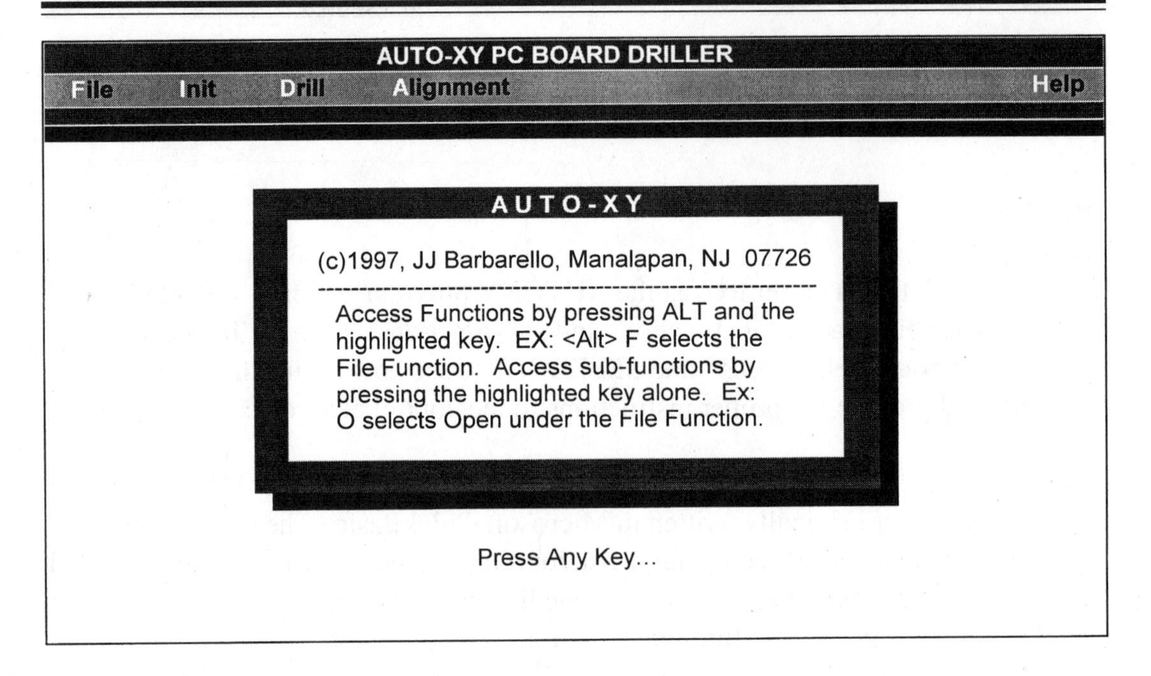

***Screen 10-1**. The opening AUTOXY program screen.*

4. Change to that new directory by typing CD AUTOXY and pressing Enter.
5. Copy all the files from the floppy disk by typing COPY A:*.* (or B:*.*) and pressing Enter.

Now, whenever you want to run the program, go to the AUTOXY directory, type AUTOXY and press Enter.

Getting Familiar with the Software

Run the AUTOXY software. You'll see the opening display of *Screen 10-1*. The information in the middle of the screen tells you the key sequences that are required to select functions and sub-functions. If you're familiar with Microsoft Windows, you'll notice that this program has a similar screen look and uses similar key sequences.

The five functions contained in the top banner are "File," "Init," "Drill," "Alignment," and "Help." The "File," "Init," and "Drill" functions also have sub-functions, which can be seen by holding down the *ALT* key and pressing the highlighted letter of the function. For instance, holding down *ALT* and pressing *F* (<Alt>F) displays the "File"

FUNCTION NAME		KEYS	WHAT IT DOES
File		<Alt>F	Selects the following File sub-functions
	Open	O	Opens (Loads) a previously created drill location file. The file name AND extension must be cited.
	Exit	X	Ends program.
	Dos Shell	D	Temporary transfer to DOS. Typing *EXIT* and pressing *Enter* returns to program.
Init		<Alt>I	Allows you to toggle (by pressing the <Tab> key) between Manual and Automatic initialization.
	Manual		Table and drill are immediately initialized (returned to their 0,0 position). Initialization will NOT occur prior to start of manual or auto drilling.
	Auto		Initialization will occur prior to auto drilling only.
Drill		<Alt>D	Selects the following Drill sub-functions.
	Manual	M	Allows you to move to a specific x-y location. No initialization occurs.
	Auto	A	Initializes to 0,0 if Auto Init has been selected, and performs the drilling sequence for the previously opened file.
Alignment		<Alt>A	Initiates the Alignment function, allowing you complete real-time control of the table and drill stepper motors.
Help		<Alt>H	Allows you to view the Help file.

***Table 10-1**. Summary of functions and sub-functions.*

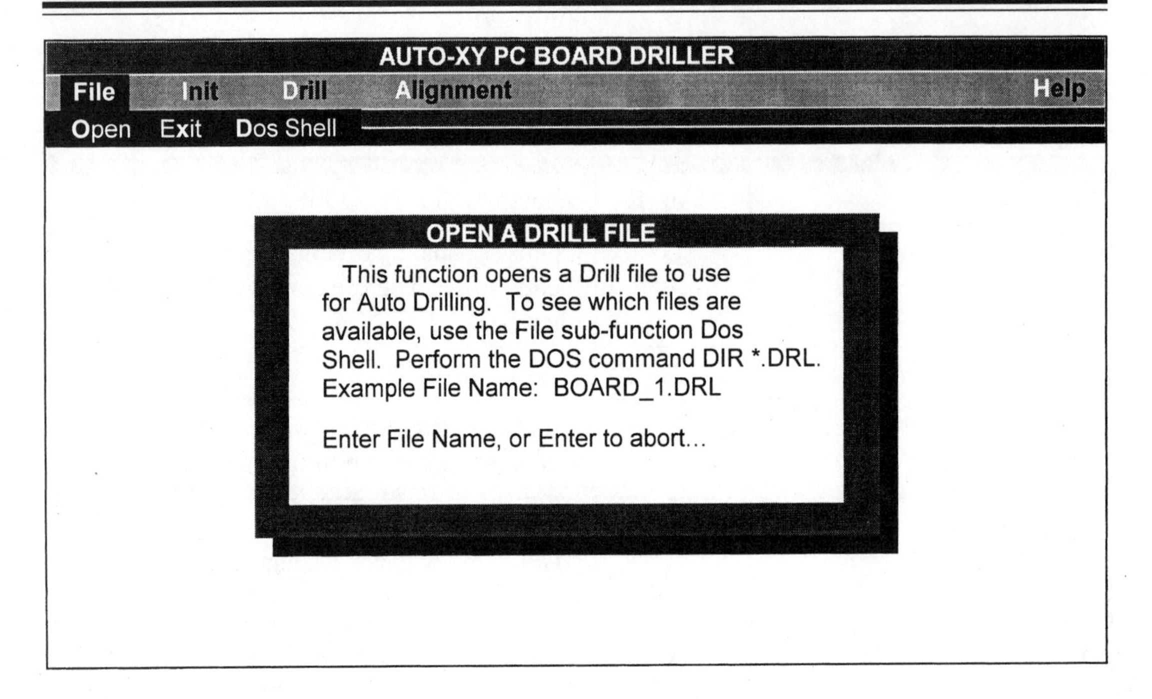

Screen 10-2*. Message box that appears after selection of the Open sub-function.*

sub-functions. The sub-functions under "File," "Init," and "Drill" provide you with the ability to load previously created drill files, switch between manual and automatic initialization of the *Auto-XY* hardware, manually or automatically drill a PC board, switch temporarily to DOS to display a directory of files, and of course exit (end) the program. All functions and sub-functions are summarized in *Table 10-1*. Let's now take a look at each of the functions and sub-functions.

File Sub-Function "DOS Shell"

The DOS shell sub-function allows you to temporarily enter DOS to perform such functions as obtaining a directory listing of all drill (*.DRL) files. Try the DOS shell sub-function by performing the following:

1. Press *<Alt>F* to select the "File" sub-functions.
2. Press D to select the DOS shell sub-function. You'll see as standard DOS screen.
3. Type *DIR *.DRL* and press Enter. The files with the extension DRL will be displayed.
4. Type Exit to return to the program and the functions.

File Sub-Function "Exit"

To end the program, press *<Alt>F* to select the "File" sub-functions, and then press *X* to select the "Exit" sub-function. The screen will clear and the message "SESSION ENDED" will appear.

File Sub-Function "Open"

The "Open" sub-function (selected by pressing O) allows you to open a previously created drill file. That file's contents then become active for any automatic drilling you choose to do until another file is opened.

Screen 10-2 shows the message box that appears when you select the Open sub-function. If you've selected this sub-function by mistake, simply press Enter to return to the functions. Otherwise, type in the full name of the file you want to load.

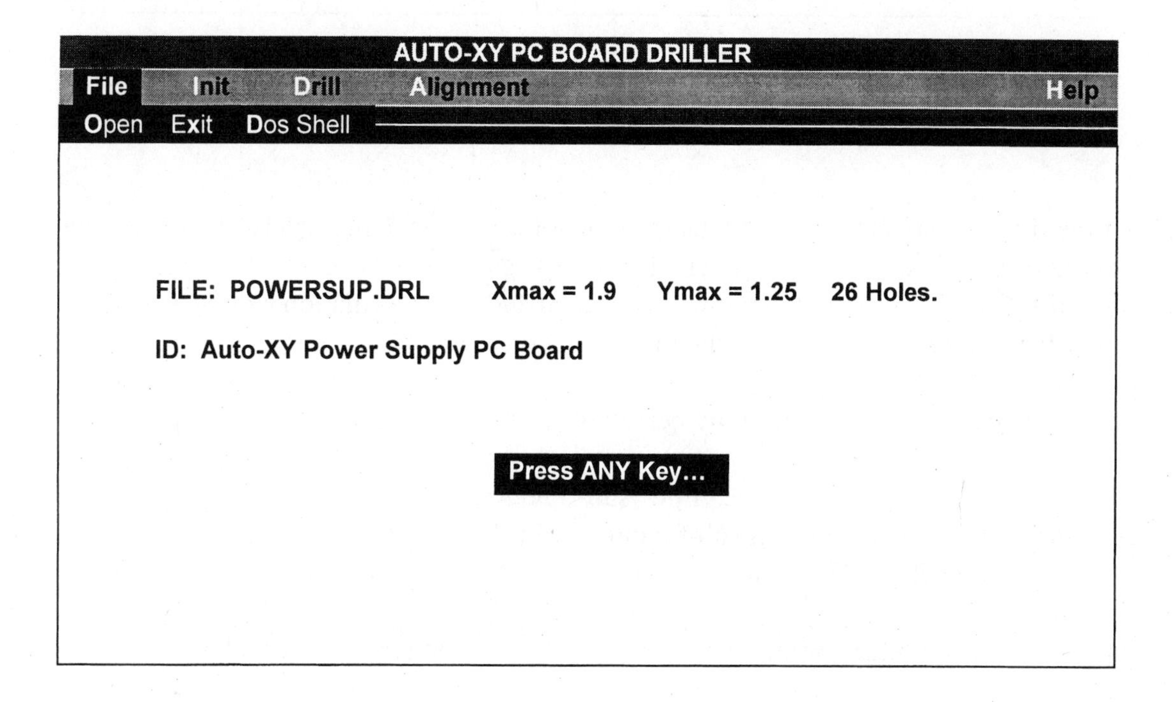

Screen 10-3. *The POWERSUP.DRL screen.*

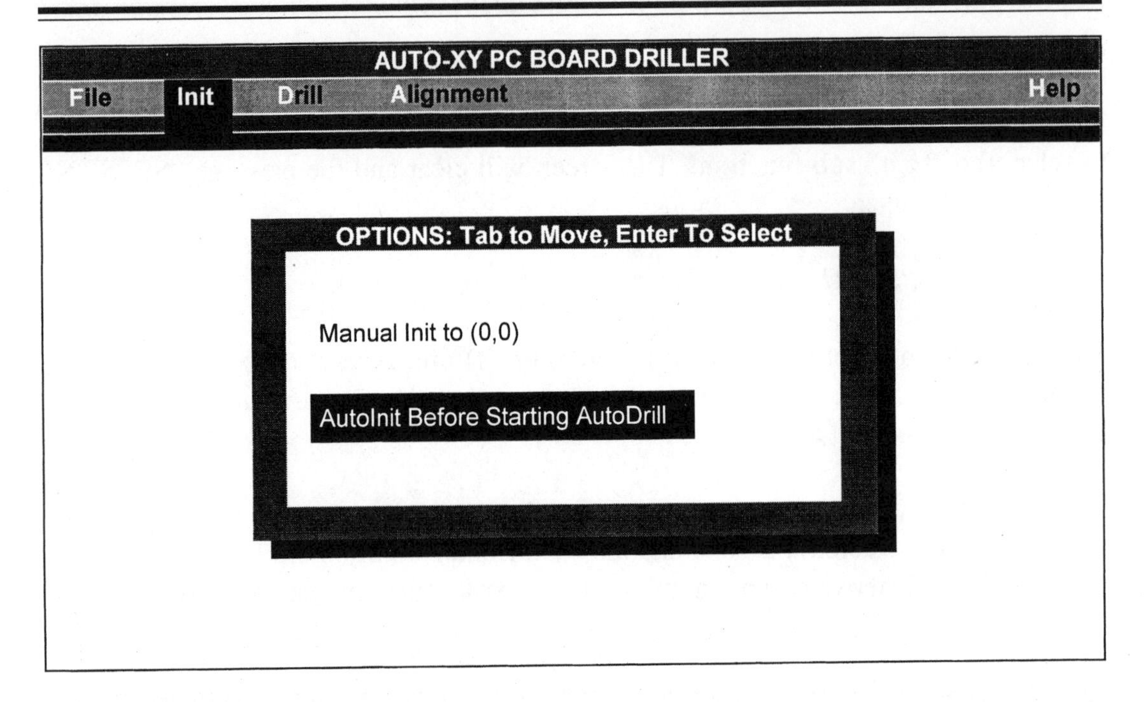

Screen 10-4. *The Init function screen.*

If the file does not exist, you will be informed that the file name you typed in does not exist. You will be asked to press any key to try again, or press Escape to return to the functions. If the file does exist, it will be loaded and file information will be displayed. Pressing any key returns you to the main function level.

Let's try the Open sub-function by performing the following:

1. Press *<ALT-F>* to select the "File" sub-functions.
2. Press O to select the Open sub-function. You'll see the display of *Screen 10-2*.
3. Type POWERSUP.DRL and press *Enter*. After a few seconds, you'll see *Screen 10-3*.
4. Press the spacebar (or any other key) to return to the main function level.

Init Function

This function allows you to toggle between its two options, Manual Init to (0,0) and AutoInit before starting AutoDrill. Manual Init immediately moves the drill and then the table until they trip their limit switches (which positions them at their zero location). No further initialization will occur.

AutoInit tells the program that you want to initialize each time you select the AutoDrill function. AutoInit does not perform any initialization during manual drilling. Let's try out the init functions by performing the following:

1. Press *<Alt>I* to select the "Init" function. You'll see *Screen 10-4*.
2. Note that pressing <Tab> toggles the highlight (now currently on the AutoInit line) between the two options.
3. Press <Tab> until the AutoInit option is selected. Then press Enter.

Drill Function

This function contains two sub-functions, Manual and AutoDrill. AutoDrill only operates if a file has been previously opened. In addition, AutoDrill will not automatically initialize the hardware to the (0,0) position if the Manual Init option is active. Manual Drill allows you to specify any X,Y location and move the hardware to that location.

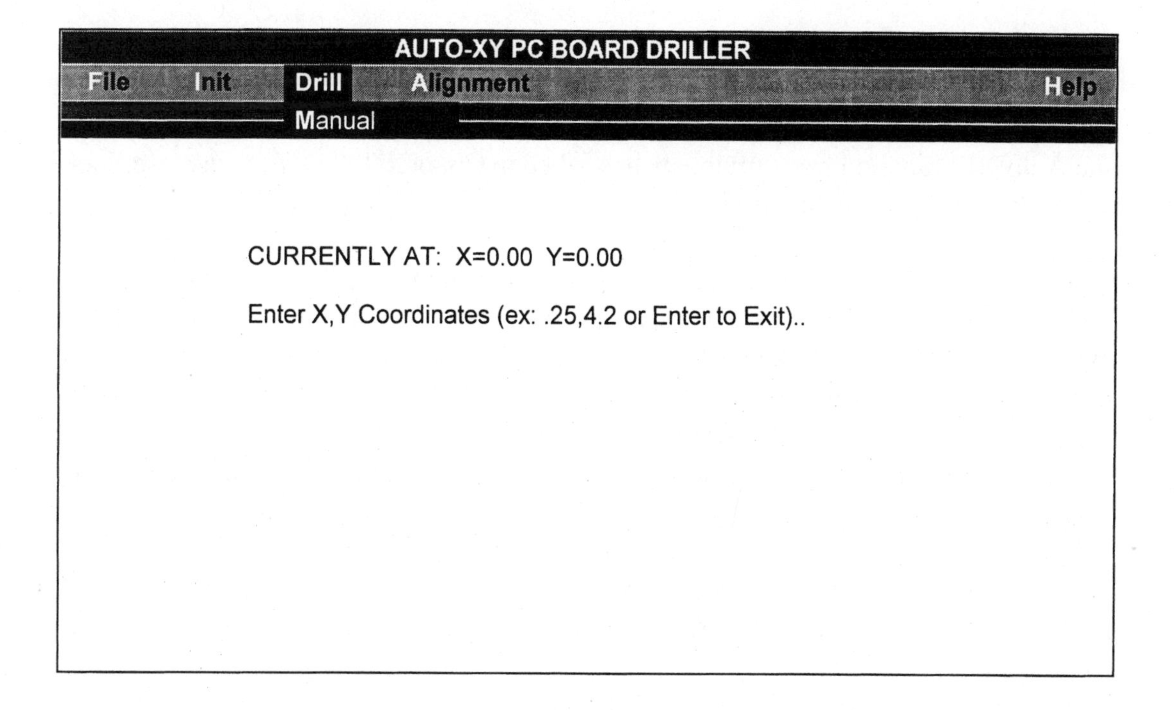

***Screen 10-5**. The Manual Drill sub-option screen.*

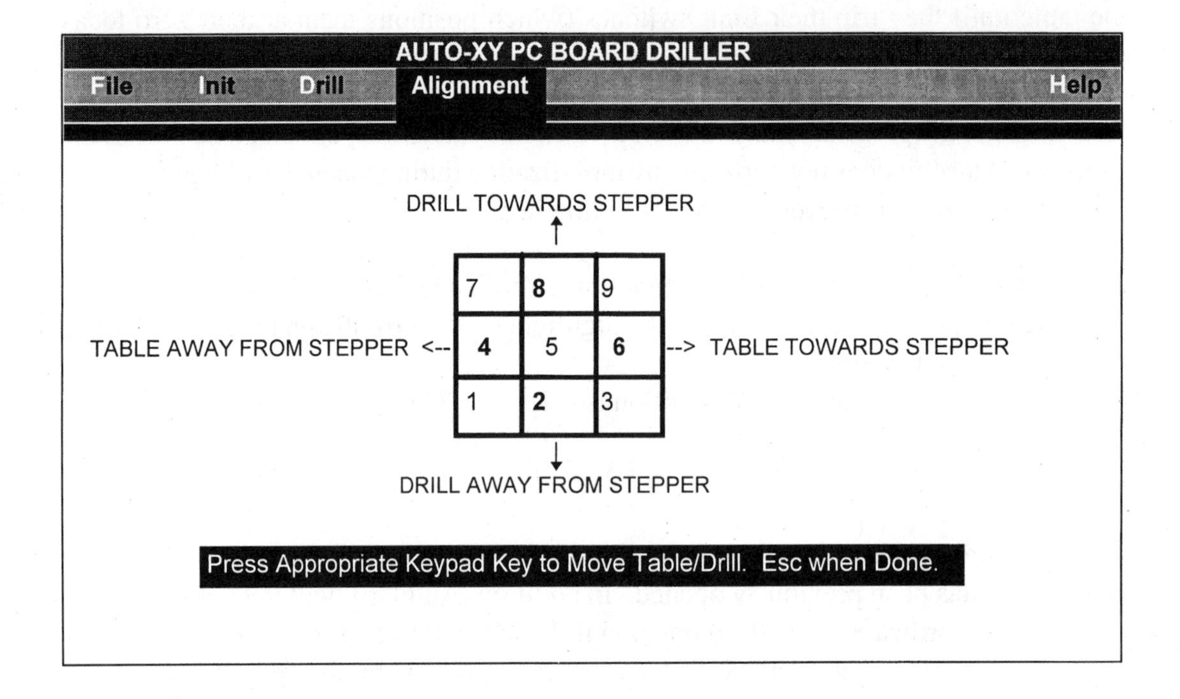

Screen 10-6. *Alignment keypad screen.*

The AutoDrill function was previously described in Chapter 9. Let's take a closer look at the Manual Drill sub-function.

Screen 10-5 is what you'll see when you select the Manual Drill sub-option. The top line shows the current location, and the next line asks for your input. You should be aware that AUTOXY only keeps track of location during manual and auto drilling. It does not know if your hardware is initialized when you first start the program. Similarly, it does not track position during movement with the Alignment function. With this in mind, it is advisable to manually initialize the hardware before beginning to use Manual Drill.

Entering any X,Y coordinates will cause the hardware to immediately move to that position. To exit this sub-function, simply press Enter without entering any coordinates.

Alignment

This function allows you to directly control movement of both the table and drill using the keypad number keys, as shown in *Screen 10-6*. As with all other movement functions, movement will stop when either limit switch is encountered. However, movement will not stop in the direction opposite movement towards the limit switches. For this reason, you must watch to ensure that the table does not become disconnected from its driver rod, and the drill does not jam against the drill bridge.

Also, as indicated above, position is not tracked during alignment. Accordingly, if you subsequently use either drill function, you should first initialize the hardware.

"Help" Function

This function (called by pressing <*Alt*>*H*) calls up and displays the DRLHELP.DAT help file. You can scroll through the help file using either the up and down arrow keys, or the <PgUp> and <PgDn> keys. To exit the Help function, press the <Esc> key.

Finding a Parallel Port

Program Listing 10-1 shows the program code that allows your PC to find any of four available parallel ports in a standard PC configuration:

```
REM**  FINDPORT.BAS
REM**   v970222, (c) 1997 James J Barbarello, Manalapan, NJ
07726
REM**
CLS : DEF SEG = 64
LOCATE 1, 25: PRINT "FIND AVAILABLE PARALLEL PORT(s)"
LOCATE 2, 1: PRINT STRING$(80, 220);
FOR PNO = 1 TO 4
addr = PEEK(PNO * 2 + 6) + 256 * (PEEK(PNO * 2 + 7))
LOCATE 6 + PNO * 2, 14
PRINT USING "PARALLEL PORT #:    "; PNO;
IF addr = 0 THEN
PRINT "Not Found"
ELSE
OUT addr, 85
IF INP(addr) = 85 THEN
PRINT "@ Address"; addr; "Decimal  ";
```

```
PRINT "("; HEX$(addr); " Hex)"
ELSE
PRINT "Not Responding"
END IF
END IF
NEXT PNO
LOCATE 18, 29: PRINT "Press ANY key to end...";
a$ = INPUT$(1)
END
```

Let's take a look at how to find and verify a port addresses:

```
CLS : DEF SEG = 64
. . .

FOR PNO = 1 TO 4
 addr = PEEK(PNO * 2 + 6) + 256 * (PEEK(PNO * 2 + 7))
. . .
```

The first line defines the memory segment which holds the registers containing the addresses of the parallel ports. This line must always be executed before initiating any communication with any parallel ports.

The next two lines loop through the four possible parallel ports (PNO) and find their addresses using the formula shown above for the variable *addr*. As an example, for parallel port number 1 (PNO = 1), its address is obtained by looking at (PEEKing) memory locations 8 (1 * 2 + 6) and 9 (1 * 2 + 7). The least significant bit (memory location 8) is added to the most significant bit (memory location 9) multiplied by 256. Similarly, port number 2 has its address stored at memory locations 10 and 11. If the contents of the memory locations result in an address of zero, there is no parallel port installed. A non-zero result indicates a physical parallel port.

Once the locations have been identified, it's important to test those locations to ensure the ports are actually working:

```
. . .
OUT addr, 85
IF INP(addr) = 85 THEN
PRINT "@ Address"; addr; "Decimal  ";
PRINT "("; HEX$(addr); " Hex)"
```

```
ELSE
PRINT "Not Responding"
END IF
. . .
```

The first line outputs the decimal number 85 to a port. In binary form, 85 is represented as 01010101. This arbitrary bit pattern is sent to the parallel port. Then, the *IF* statements read back the contents of the port and compare it to that pattern. If the patterns sent and received are the same, the port is responding properly and the next two lines print the port's address (in both decimal and hexadecimal format). If the port does not return the arbitrary bit pattern, the port is presumed to be not responding, and that result is reported in the next print line.

Communicating with the Control Electronics

Program Listing 10-2 shows the program code for the SPEEDSET program:

```
REM:   Speed Set
REM:   SPEEDSET.BAS
REM:   V970222
CLS : CLEAR : DEF SEG = 64: DEFINT A
DIM a(4)
a(1) = 5: a(2) = 3: a(3) = 10: a(4) = 12
LOCATE 1, 26: PRINT "SET SPEED OF STEPPER MOTORS"
LOCATE 2, 1: PRINT STRING$(80, 220);
'
'GET SETUP DATA IN DRLSETUP.DAT FILE
'
CLOSE 1: OPEN "DRLSETUP.DAT" FOR INPUT AS #1
LINE INPUT #1, add$: add = VAL(add$)
LINE INPUT #1, Speed$: Speed! = VAL(Speed$)
LINE INPUT #1, distancecal$: distancecal = VAL(distancecal$)
LINE INPUT #1, messagedisplay$
CLOSE 1
OUT add, a(4) + 128: FOR i = 1 TO Speed!: NEXT i: OUT add, 128
OUT add, a(4): FOR i = 1 TO Speed!: NEXT i: OUT add, 0: aseq = 4
'
'DO SPEED SET
'
speedset:
```

```
COLOR 7, 0: VIEW PRINT 6 TO 24: CLS : VIEW PRINT
LOCATE 6, 22: PRINT "Current Speed Factor:"; Speed
LOCATE 8, 24: PRINT "PRESS     TO"
LOCATE 9, 24: PRINT "—     ————————————————"
LOCATE 10, 26: PRINT "2      Move Drill Stepper Clockwise"
LOCATE 11, 26: PRINT "4      Move Table Stepper Clockwise"
LOCATE 12, 26: PRINT "6      Move Table Stepper Counterclock-
wise"
LOCATE 13, 26: PRINT "8      Move Drill Stepper Counterclock-
wise"
LOCATE 14, 25: PRINT "Esc     End"
LOCATE 15, 24: PRINT "Enter     Change Speed Factor"
speedset01:
a$ = INKEY$: IF a$ = "" THEN GOTO speedset01
adir = ASC(RIGHT$(a$, 1))
SELECT CASE adir
CASE IS = 13
GOSUB changefactor: GOTO speedset
CASE IS = 27: 'Esc
OUT add, 0 + which: GOTO newdata
CASE IS = 50, 80: 'Drill Away From Stepper - 2
which = 0: steps = 1
CASE IS = 52, 75: 'Table Away From Stepper - 4
which = 128: steps = 1
CASE IS = 54, 77: 'Table Towards Stepper - 6
which = 128: steps = -1
CASE IS = 56, 72: 'Drill Towards From Stepper - 8
which = 0: steps = -1
CASE ELSE
BEEP: GOTO speedset
END SELECT
COLOR 23, 0: LOCATE 20, 28: PRINT " Press Any Key to Stop."
OUT add, which: stemp! = Speed!
speedset02:
aseq = (aseq MOD 4) + steps
IF aseq = 0 THEN aseq = 4
IF aseq = -1 THEN aseq = 3
OUT add, a(aseq) + which
FOR k = 1 TO stemp!: NEXT k
IF (INP(add + 1) AND 64) = 0 AND adir = 77 AND which = 128 THEN
GOTO am3
```

```
IF (INP(add + 1) AND 128) = 128 AND adir = 80 AND which = 0 THEN
GOTO am3
a$ = INKEY$: IF a$ = "" THEN GOTO speedset02
'a = ASC(a$): IF a <> 27 THEN GOTO speedset02
am3:
start! = TIMER
WHILE (TIMER - start!) < .2: WEND
OUT add, 0 + which
GOTO speedset
changefactor:
LOCATE 18, 22: PRINT "Enter New Speed Factor (1 to 32767),"
LOCATE 19, 38: PRINT "or"
LOCATE 20, 22: PRINT "Press Enter to Keep Current Factor...";
LINE INPUT newspeed$
IF newspeed$ = "" THEN RETURN
newspeed! = VAL(newspeed$)
IF newspeed! < 1 OR newspeed! > 32767 THEN
BEEP
GOTO changefactor
END IF
Speed! = newspeed
RETURN
'
'WRITE NEW DATA TO DRLSETUP.DAT FILE
'
newdata:
VIEW PRINT 6 TO 24: CLS : VIEW PRINT
LOCATE 10, 10: PRINT "Ready to End. "
LOCATE 12, 10: PRINT "Do you Want to Save the NEW Speed Data (Y/
N)? ...";
newdata01:
a$ = INKEY$: IF a$ = "" THEN GOTO newdata01
a$ = UCASE$(a$)
SELECT CASE a$
CASE IS = "Y"
GOTO newdata02
CASE IS = "N"
LOCATE 14, 10: PRINT "No Save Selected.  Program Ended.": END
CASE ELSE
BEEP: GOTO newdata01
END SELECT
```

```
newdata02:
LOCATE 14, 10: PRINT "Saving New Data ...";
CLOSE 1: OPEN "DRLSETUP.DAT" FOR OUTPUT AS #1
add$ = LTRIM$(STR$(add))
PRINT #1, add$
Speed$ = LTRIM$(STR$(Speed!))
PRINT #1, Speed$
PRINT #1, distancecal$
PRINT #1, messagedisplay$
CLOSE 1
PRINT "Done.  Program Ended"
END
```

The process for communicating with the control electronics is the same in SPEEDSET and AUTOXY:

```
. . .
CLS : CLEAR : DEF SEG = 64: DEFINT A
DIM a(4)
a(1) = 5: a(2) = 3: a(3) = 10: a(4) = 12
. . .
```

Note that the program begins with the standard communication mechanism (DEF SEG = 64). The next two lines dimension a four element array (a(4)) and then populate it with the control sequence values discussed in Chapter 2. Although most forms of Basic will automatically define an array for you the first time you specify any array element, manually defining it uses less memory (the automatic definition identifies eleven array elements, while we only need four):

```
. . .
CLOSE 1: OPEN "DRLSETUP.DAT" FOR INPUT AS #1
LINE INPUT #1, add$: add = VAL(add$)
LINE INPUT #1, Speed$: Speed! = VAL(Speed$)
LINE INPUT #1, distancecal$: distancecal = VAL(distancecal$)
LINE INPUT #1, messagedisplay$
CLOSE 1
. . .
```

The next five lines open the DRLSETUP.DAT file and retrieve the setup values. Since these data are stored as plain text, and the first three elements need to be numeric values, the VAL function converts them from text to numeric. Once the retrieval process is complete, the file is closed:

```
. . .
SELECT CASE adir
CASE IS = 13:  'Enter
GOSUB changefactor: GOTO speedset
CASE IS = 27: 'Esc
OUT add, 0 + which: GOTO newdata
CASE IS = 50, 80: 'Drill Away From Stepper - 2
which = 0: steps = 1
CASE IS = 52, 75: 'Table Away From Stepper - 4
which = 128: steps = 1
CASE IS = 54, 77: 'Table Towards Stepper - 6
which = 128: steps = -1
CASE IS = 56, 72: 'Drill Towards From Stepper - 8
which = 0: steps = -1
CASE ELSE
BEEP: GOTO speedset
END SELECT
. . .
```

The SELECT CASE routine waits for a keypress and decides what to do. A separate case is established for each of the four possible number keys (2, 4, 6, and 8), as well as pressing the Enter and Escape keys. When a key is pressed, its ASCII value is determined and stored in the variable *adir*. That ASCII value is used to select the appropriate case.

Note that for the four possible number keys, the variables *which* and *steps* are set to particular values. The variable *which* is used to determine which of the drill or table are to be controlled. The variable *steps* is used to determine if the selected motor should turn clockwise or counterclockwise.

If you skip ahead for a moment to Chapter 11 and review the control electronics schematic diagram, you will notice that parallel port pin 9 provides a control signal to the driver circuit for relay K1. If pin 9 is high, the driver circuit will energize the relay and engage motor M1 (which drives the table). If pin 9 is low, K1 will be de-energized, engaging M2 (which drives the drill). To make pin 9 high, we need to send a decimal 128 to the parallel port's base address (*add*); to make it low, we send a decimal zero.

So, whenever we are addressing the table, we make the variable *which* equal to 128. When we address the drill, we make *which* equal to zero.

Now that we've selected the appropriate motor, we have to send a control sequence that will drive it either clockwise or counterclockwise. If you refer back to Chapter 2, you will see that the decimal values of the control sequence are reflected in the *A* array. Back in Chapter 2 we also learned that sending the control sequence produces one direction of motion, and sending the sequence in the reverse order produces the opposite direction of motion. So, the variable *steps* is set to 1 when we want the motor to turn clockwise (this specifies that the control values should be sent out in the sequence stored in the *A* array). When we want counterclockwise rotation, we set the variable *steps* to -1, which will send the control values out in reverse sequence.

With these variables set, we can now proceed to the program code that actually causes motor movement. The code below begins with sending the proper control signal to the relay driver circuitry. The three lines defining *aseq* require some explanation. The first line uses the modulus (MOD) function. If *aseq* is between 1 and 4 (*aseq MOD 4*), it will not have any noticeable affect. If *aseq* = 5, the result will be 1 (since MOD 4 essentially divides the number by 4 and returns the remainder of that division). If *aseq* equals 4, the result will be 0. So, (*aseq MOD 4*) essentially loops the value between 0 and 3. If *steps* equals 1, the loop will be 1 to 4.

If *steps* equals -1, (*aseq MOD 4*) + *steps* will be (*aseq MOD 4*) *-1*. This will result in the loop -1, -2, -3, -4. If we now redefine *aseq* as 3 when it becomes -1, the loop will be converted to 3, 2, 1, 0. If we then redefine *aseq* as 4 when it becomes 0, the loop is again converted, this time to 4, 3, 2, 1. As a result, we get a loop that is 1, 2, 3, 4 when *steps* = *1* and a loop that is 4, 3, 2, 1 when *steps* = *-1*. Note the line *OUT add, a(aseq) + which*. This sends out the control values in the sequence defined by the *aseq* loop (a(1), a(2), a(3), a(4), or a(4), a(3), a(2), a(1)), thus specifying clockwise or counterclockwise rotation by simply changing the value of *steps*. The value outputted has the value of *which* added to it, to ensure the relay stays in the same state it was previously set to. The *For..Next* loop provides a delay between subsequent control sequences to allow the motor time to respond:

```
. . .
OUT add, which: stemp! = Speed!
speedset02:
aseq = (aseq MOD 4) + steps
IF aseq = 0 THEN aseq = 4
```

```
IF aseq = -1 THEN aseq = 3
OUT add, a(aseq) + which
FOR k = 1 TO stemp!: NEXT k
IF (INP(add + 1) AND 64) = 0 AND adir = 77 AND which = 128 THEN
GOTO am3
END IF
IF (INP(add + 1) AND 128) = 128 AND adir = 80 AND which = 0 THEN
GOTO am3
END IF
a$ = INKEY$: IF a$ = "" THEN GOTO speedset02
am3:
start! = TIMER
WHILE (TIMER - start!) < .2: WEND
OUT add, 0 + which
. . .
```

Now let's look at the first of the two *IF* statements. The *INP(add + 1) AND 64* returns the status of parallel port pin 10, which is connected to the table limit switch S1. If the value returned is zero, the switch is closed. Further, if *which = 128* the table stepper motor is engaged. Finally, if *adir = 77* (previously set in the SELECT CASE routine), the table is moving towards the stepper. These three conditions indicate that the table stepper motor has been engaged and moving the Table towards it, and the table limit switch has been tripped. In that event, the *IF* statement passes execution to *am3:* which waits a short while and turns off the motor (*OUT add, 0 + which*) while keeping the relay in its previously set state.

Similarly, the second *IF* statement checks the status of parallel port pin 11 *(INP(add + 1 and 128))*. If the value returned is 128, drill limit switch S2 is closed. Further, if *which = 0*, the relay is de-energized and the drill stepper motor is engaged. Finally, if *adir = 80*, the drill is moving away from its stepper. These three conditions indicate that the drill stepper motor has been engaged and moving the drill away from it, and the drill limit switch has been tripped. As before, the *IF* statement passes execution to *am3:* .

The AUTOXY Program

Program Listing 10-3 lists the full program code for AUTOXY:

```
REM:   Auto-XY
REM:   AUTOXY.BAS
```

```
REM:  V970222 - Auto-XY Software with Lashback Correction
REM:
CLS : CLEAR : ON ERROR GOTO errortrap: DEF SEG = 64: DEFINT A
DIM a(4), help$(1000), initmode$(3), x(500), y(500), z(500)
a(1) = 5: a(2) = 3: a(3) = 10: a(4) = 12
initmode$(1) = "Manual Init to (0,0)"
initmode$(2) = "AutoInit Before Starting AutoDrill": initmode =
2
GOSUB GetSetupAndInitSteppers
'
'INITIAL SCREEN SETUP ***********************
'
COLOR 7, 1: CLS : VIEW PRINT 2 TO 4: COLOR 0, 3: CLS
VIEW PRINT 4 TO 25: COLOR 0, 7: CLS : VIEW PRINT: COLOR 7, 1
LOCATE 1, 28: PRINT "AUTO-XY PC BOARD DRILLER": GOSUB commandline
PCOPY 0, 1: IF messagedisplay$ <> "NO" THEN GOSUB startupmessage
PCOPY 1, 0
'
'SELECT COMMAND LINE OPTIONS/SUB-OPTIONS ************************
'
getkey:
a$ = INKEY$: IF a$ = "" THEN GOTO getkey
a = ASC(RIGHT$(a$, 1))
SELECT CASE a
CASE IS = 23: ' INIT
LOCATE 2, 9: COLOR 15, 0: PRINT " Init ";
LOCATE 3, 9: PRINT SPACE$(6); : GOSUB initmode
IF initmode = 1 THEN GOSUB zero
CASE IS = 30
LOCATE 2, 26: COLOR 15, 0: PRINT " Alignment ";
LOCATE 3, 26: PRINT SPACE$(11)
GOSUB alignment
CASE IS = 32: 'DRILL
GOSUB selectdrill: opt$ = "MA" + CHR$(27): GOSUB getoption
SELECT CASE a$
CASE IS = "M"
LOCATE 3, 25: PRINT SPACE$(5); : GOSUB manualdrill
CASE IS = "A"
LOCATE 3, 18: PRINT SPACE$(6): GOSUB autodrill
CASE ELSE
END SELECT
CASE IS = 33: 'FILE
```

```
GOSUB selectfile: opt$ = "DOX" + CHR$(27): GOSUB getoption
SELECT CASE a$
CASE IS = "X"
CLS : LOCATE 18, 1: PRINT "SESSION ENDED.": END
CASE IS = "O"
GOSUB openafile
CASE IS = "D"
PCOPY 0, 1
COLOR 7, 0: CLS
SHELL
PCOPY 1, 0: GOSUB commandline
CASE ELSE
END SELECT
CASE IS = 35: 'HELP
GOSUB helpsub
CASE ELSE
SOUND 800, 1: SOUND 500, 1
END SELECT
GOSUB commandline: GOTO getkey
'
'END OF MAIN PROGRAM - SUBROUTINES FOLLOW
'
'SUBROUTINES **********************************
'
commandline:
LOCATE 2, 1: COLOR 0, 3: PRINT SPACE$(80);
COLOR 0, 1: LOCATE 3, 1: PRINT STRING$(80, 220); : COLOR 7, 1
LOCATE 2, 3: COLOR 4, 3: PRINT "F"; : COLOR 0, 3: PRINT "ile"
LOCATE 2, 10: COLOR 4, 3: PRINT "I"; : COLOR 0, 3: PRINT "nit"
LOCATE 2, 18: COLOR 4, 3: PRINT "D"; : COLOR 0, 3: PRINT "rill"
LOCATE 2, 27: COLOR 4, 3: PRINT "A"; : COLOR 0, 3: PRINT "lignment"
LOCATE 2, 75: COLOR 4, 3: PRINT "H"; : COLOR 0, 3: PRINT "elp"
RETURN
'
optionbox:
COLOR 0, 7: FOR i = 7 TO 19: LOCATE i, 14: PRINT STRING$(54, 32):
NEXT
COLOR 1, 7: LOCATE 8, 15: PRINT STRING$(51, 219)
COLOR 7, 1: LOCATE 8, 22: PRINT "OPTIONS: Tab to Move, Enter to
Select";
COLOR 1, 7: LOCATE 18, 15: PRINT STRING$(51, 219)
FOR i = 9 TO 17
```

```
LOCATE i, 15: PRINT STRING$(2, 219)
LOCATE i, 64: PRINT STRING$(2, 219)
NEXT i
COLOR 0, 7: LOCATE 8, 66: PRINT CHR$(220)
LOCATE 19, 16: PRINT STRING$(51, 223)
FOR i = 9 TO 18
LOCATE i, 66: PRINT CHR$(219)
NEXT i
RETURN
'
'FILE COMMAND OPTIONS
'
selectfile:
LOCATE 2, 2: COLOR 15, 0: PRINT " File ";
LOCATE 3, 1
COLOR 0, 1: PRINT STRING$(80, 220);
LOCATE 3, 2: COLOR 12, 0: PRINT " O"; : COLOR 7, 0: PRINT "pen
E";
COLOR 12, 0: PRINT "x"; : COLOR 7, 0: PRINT "it  ";
COLOR 12, 0: PRINT "D"; : COLOR 7, 0: PRINT "os Shell ";
RETURN
'
'DRILL COMMAND OPTIONS
'
selectdrill:
LOCATE 2, 17: COLOR 15, 0: PRINT " Drill ";
LOCATE 3, 1
COLOR 0, 1: PRINT STRING$(80, 220);
LOCATE 3, 17: COLOR 12, 0: PRINT " M"; : COLOR 7, 0: PRINT "anual
";
COLOR 12, 0: PRINT "A"; : COLOR 7, 0: PRINT "uto ";
RETURN
'
'CHECK SUB-OPTIONS' VALIDITY
'
getoption:
a$ = UCASE$(INPUT$(1))
IF INSTR(opt$, a$) = 0 THEN
SOUND 800, 1: SOUND 200, 1: GOTO getoption
END IF
RETURN
'
```

```
'SUB-OPTION OPEN FILE
'
openafile:
PCOPY 0, 1
openafile1:
GOSUB optionbox
COLOR 1, 7: LOCATE 8, 15: PRINT STRING$(51, 219)
COLOR 7, 1: LOCATE 8, 32: PRINT "OPEN A DRILL FILE": COLOR 0, 7
LOCATE 9, 19: PRINT "   This function opens a Drill file to use"
LOCATE 10, 19: PRINT "for Auto Drilling.  To see which files are"
LOCATE 11, 19: PRINT "available, use the File sub-function Dos"
LOCATE 12, 19: PRINT "Shell. Perform the DOS command DIR *.DRL."
LOCATE 13, 19: PRINT "Example File Name:  BOARD_1.DRL"
LOCATE 15, 19: PRINT "Enter File Name, or Enter to abort.";
LOCATE 16, 30: LINE INPUT drillfile$: LF = 0
IF LEN(drillfile$) = 0 THEN PCOPY 1, 0: RETURN
CLOSE 1: OPEN "r", 1, drillfile$: FIELD 1, 1 AS df$: LF = LOF(1)
IF LF = 0 THEN
CLOSE : KILL drillfile$
LOCATE 15, 19: PRINT SPACE$(40)
LOCATE 15, 19: PRINT "File "; drillfile$; " not Found."
LOCATE 16, 19: PRINT "Press ANY key to try again, Esc to return"
drillfileloop:
a$ = INKEY$: IF a$ = "" THEN GOTO drillfileloop
a = ASC(a$)
SELECT CASE a
CASE IS = 27
PCOPY 1, 0
CLOSE : drillfile$ = ""
RETURN
CASE ELSE
GOTO openafile1
END SELECT
END IF
LOCATE 15, 19: PRINT SPACE$(40)
LOCATE 16, 19: PRINT SPACE$(40)
LOCATE 15, 19: PRINT "Reading File "; drillfile$; "...";
CLOSE 1: OPEN drillfile$ FOR INPUT AS #1
LINE INPUT #1, title$
LINE INPUT #1, boardsize$
xmax = VAL(boardsize$)
comma = INSTR(boardsize$, ",")
```

```
ymax = VAL(MID$(boardsize$, comma + 1, 20))
ctr1 = 1
WHILE NOT EOF(1)
LINE INPUT #1, a$
x(ctr1) = VAL(a$)
comma1 = INSTR(a$, ",")
y(ctr1) = VAL(MID$(a$, comma1 + 1, 20))
comma2 = INSTR(comma1 + 1, a$, ",")
z(ctr1) = VAL(MID$(a$, comma2 + 1, 20))
ctr1 = ctr1 + 1
WEND
numberholes = ctr1 - 1
PRINT "Done."
PCOPY 1, 0
LOCATE 10, 10: PRINT "FILE: "; UCASE$(drillfile$)
LOCATE 10, 36: PRINT "Xmax ="; xmax; "  Ymax ="; ymax;
PRINT " "; numberholes; "Holes."
LOCATE 12, 10: PRINT "ID: "; title$
COLOR 15, 0: LOCATE 16, 33: PRINT " Press ANY key... "; : COLOR
0, 7
a$ = INPUT$(1)
PCOPY 1, 0
RETURN
'
'INIT MODE SUB-OPTIONS
'
initmode:
PCOPY 0, 1
GOSUB optionbox
suboption = initmode
initmodeloop01:
FOR i = 12 TO 14 STEP 2
IF suboption = ((i - 10) / 2) THEN COLOR 12, 0 ELSE COLOR 0, 7
LOCATE i, 20: PRINT initmode$((i - 10) / 2)
NEXT i
initmodeloop02:
aa$ = INKEY$: IF aa$ = "" THEN GOTO initmodeloop02
aa = ASC(aa$)
SELECT CASE aa
CASE IS = 9
suboption = suboption + 1
IF initmode$(suboption) = "" THEN suboption = 1
```

```
CASE IS = 13
initmode = suboption: PCOPY 1, 0: RETURN
CASE IS = 27
PCOPY 1, 0: RETURN
CASE ELSE
SOUND 800, 1: SOUND 500, 1
END SELECT
GOTO initmodeloop01
'
'HELP OPTION
'
helpsub:
PCOPY 0, 1
COLOR 0, 7: VIEW PRINT 4 TO 25: CLS
CLOSE : ERASE help$
LOCATE 13, 30: PRINT "Retrieving Help File...";
CLOSE 1: OPEN "drlhelp.dat" FOR INPUT AS #1
j = 1
WHILE NOT EOF(1)
LINE INPUT #1, help$(j)
help$(j) = LEFT$(help$(j) + STRING$(80, 32), 75)
j = j + 1
WEND: CLOSE
COLOR 8, 7
FOR i = 4 TO 25: LOCATE i, 78: PRINT STRING$(3, 219); : NEXT
COLOR 7, 0
LOCATE 4, 79: PRINT CHR$(24); : LOCATE 5, 79: PRINT CHR$(178);
LOCATE 25, 79: PRINT CHR$(25);
helpsize = j - 1: increment = helpsize / 21: firstline = 1
COLOR 0, 7
helpsub1:
pointer = INT(firstline / increment)
IF pointer < 0 THEN
pointer = 0
ELSEIF pointer > 19 THEN
pointer = 19
END IF
FOR i = 5 TO 24: LOCATE i, 79: PRINT CHR$(176); : NEXT
LOCATE pointer + 5, 79: PRINT CHR$(178);
FOR i = 4 TO 25
LOCATE i, 4: PRINT help$(firstline - 4 + i);
NEXT i
```

```
helploop:
a$ = INKEY$: IF a$ = "" THEN GOTO helploop
a = ASC(RIGHT$(a$, 1))
SELECT CASE a
CASE IS = 27
GOTO nomorehelp
CASE IS = 72
IF firstline > 1 THEN firstline = firstline - 1 ELSE BEEP
CASE IS = 73
IF firstline > 20 THEN firstline = firstline - 20 ELSE BEEP
CASE IS = 80
IF firstline < (helpsize - 21) THEN
firstline = firstline + 1
ELSE
BEEP
END IF
CASE IS = 81
IF firstline < (helpsize - 41) THEN
firstline = firstline + 20
ELSE
firstline = helpsize - 21: BEEP
END IF
CASE ELSE
BEEP: GOTO helploop
END SELECT
GOTO helpsub1
nomorehelp:
ERASE help$: VIEW PRINT: PCOPY 1, 0: RETURN
'
'GET INITIALIZATION DATA FROM FILES AND INITIALIZE STEPPERS
'
GetSetupAndInitSteppers:
CLOSE 1: OPEN "DRLSETUP.DAT" FOR INPUT AS #1
LINE INPUT #1, add$: add = VAL(add$)
LINE INPUT #1, speed$: speed! = VAL(speed$)
LINE INPUT #1, distancecal$: distancecal = VAL(distancecal$)
LINE INPUT #1, messagedisplay$
messagedisplay$ = UCASE$(LEFT$(messagedisplay$, 2))
CLOSE 1
OUT add, a(4) + 128: FOR i = 1 TO speed!: NEXT i: OUT add, 128
OUT add, a(4): FOR i = 1 TO speed!: NEXT i: OUT add, 0: aseq = 4
RETURN
```

```
'
'MANUAL INITIALIZATION SUB-OPTION (Move To (0,0))
'
zero:
which = 0: steps = 1
zero01:
OUT add, which: stepcount = 0
zero02:
aseq = (aseq MOD 4) + steps
IF aseq = 0 THEN aseq = 4
IF aseq = -1 THEN aseq = 3
IF (INP(add + 1) AND 64) = 0 AND which = 128 THEN GOTO donezero
IF (INP(add + 1) AND 128) = 128 AND which = 0 THEN GOTO donezero
OUT add, a(aseq) + which
FOR k = 1 TO speed!: NEXT k
stepcount = stepcount + 1
GOTO zero02
donezero:
start! = TIMER
WHILE (TIMER - start!) < .2: WEND
OUT add, 0 + which
IF which = 0 THEN which = 128: steps = -1: GOTO zero01
ylocation = 0: xlocation = 0
RETURN
'
'MANUAL DRILLING SUB-OPTION
'
manualdrill:
VIEW PRINT 4 TO 25: COLOR 0, 7: CLS : PCOPY 0, 1
manualdrill01:
PCOPY 1, 0
LOCATE 8, 10
PRINT USING "CURRENTLY AT: X=#.##  Y=#.##"; xlocation; ylocation;
LOCATE 10, 10
PRINT "Enter X,Y Coordinates (ex: .25,4.2 or Enter to Exit).. ";
LINE INPUT a$
IF a$ = "" THEN PCOPY 1, 0: VIEW PRINT: RETURN
comma1 = INSTR(a$, ",")
IF comma1 = 0 THEN BEEP: GOTO manualdrill01
xdest = VAL(a$): ydest = VAL(MID$(a$, comma1 + 1, 20))
IF xdest < 0 OR ydest < 0 THEN BEEP: GOTO manualdrill01
CLS : LOCATE 10, 10: PRINT USING "Moving to (#.##,#.##)"; xdest;
```

```
ydest;
LOCATE 18, 33: COLOR 8, 7: PRINT "Esc To Abort. "; : COLOR 0, 7
which = 128: 'Table Selected - X Direction
xsteps = (xdest - xlocation) * distancecal
ysteps = (ydest - ylocation) * distancecal
IF xsteps < 0 THEN
steps = -1
IF xdest > .1 THEN
xsteps = ABS(xsteps) + 43
lashflgx = 1
ELSE
xsteps = ABS(xsteps)
lashflgx = 0
END IF
ELSE
steps = 1
lashflgx = 0
END IF
IF xsteps = 0 THEN GOTO donemanual
manualmove01:
OUT add, which: stepcount = 0
manualmove02:
aseq = (aseq MOD 4) + steps
IF aseq = 0 THEN aseq = 4
IF aseq = -1 THEN aseq = 3
OUT add, a(aseq) + which
FOR k = 1 TO speed!: NEXT k
stepcount = stepcount + 1
IF (stepcount >= xsteps) AND which = 128 THEN GOTO donemanual
IF (stepcount >= ysteps) AND which = 0 THEN GOTO donemanual
panicstop$ = INKEY$: IF panicstop$ = "" THEN GOTO manualmove02
IF ASC(panicstop$) = 27 THEN
PCOPY 1, 0: VIEW PRINT: OUT add, 0: OUT add, 128: RETURN
ELSE
GOTO manualmove02
END IF
donemanual:
IF lashflgx = 1 THEN
steps = 1
xsteps = 43
lashflgx = 0
GOTO manualmove01
```

```
END IF
IF lashflgy = 1 THEN
steps = -1
ysteps = 43
lashflgy = 0
GOTO manualmove01
END IF
start! = TIMER
WHILE (TIMER - start!) < .2: WEND
OUT add, 0 + which
IF which = 128 THEN
which = 0: 'Drill Selected - Y Direction
IF ysteps < 0 THEN
steps = 1:
IF ydest > .1 THEN
ysteps = ABS(ysteps) + 43
lashflgy = 1
ELSE
ysteps = ABS(ysteps)
lashflgy = 0
END IF
ELSE steps = -1
END IF
IF ysteps <> 0 THEN GOTO manualmove01
END IF
xlocation = xdest: ylocation = ydest
GOTO manualdrill01
'
'AUTO DRILLING SUB-OPTION
'
autodrill:
VIEW PRINT 4 TO 25: COLOR 0, 7: CLS : PCOPY 0, 1
currenthole = 0
IF drillfile$ = "" THEN
SOUND 800, 1: SOUND 500, 1
LOCATE 10, 20
PRINT "NO FILE OPENED.  Use File Option To Open A File."
LOCATE 12, 20: PRINT "Press Any Key..";
a$ = INPUT$(1): PCOPY 1, 0: VIEW PRINT: RETURN
END IF
COLOR 8, 7: LOCATE 6, 10: PRINT "FILE: "; UCASE$(drillfile$)
LOCATE 6, 36: PRINT "Xmax ="; xmax; "  Ymax ="; ymax;
```

```
PRINT " "; numberholes; "Holes."
LOCATE 7, 10: PRINT "ID: "; title$: COLOR 0, 7
IF initmode = 2 THEN
LOCATE 10, 10: PRINT "Initializing to (0,0).  Wait..";
GOSUB zero
PCOPY 1, 0
END IF
autodrill00:
PCOPY 1, 0
COLOR 8, 7: LOCATE 6, 10: PRINT "FILE: "; UCASE$(drillfile$)
LOCATE 6, 36: PRINT "Xmax ="; xmax; "  Ymax ="; ymax;
PRINT " "; numberholes; "Holes."
LOCATE 7, 10: PRINT "ID: "; title$: COLOR 0, 7: LOCATE 10, 10
PRINT USING "Start At Which Hole (# to ###), or Enter to Abort..";
currenthole + 1; numberholes;
LINE INPUT a$: IF LEN(a$) = 0 THEN GOTO doneauto01
wheretostart = VAL(a$)
IF wheretostart < 1 OR wheretostart > numberholes THEN
SOUND 800, 1: SOUND 500, 1
GOTO autodrill00
ELSE
currenthole = wheretostart - 1
END IF
autodrill01:
WHILE currenthole < numberholes
currenthole = currenthole + 1
PCOPY 1, 0
COLOR 8, 7: LOCATE 6, 10: PRINT "FILE: "; UCASE$(drillfile$)
LOCATE 6, 36: PRINT "Xmax ="; xmax; "  Ymax ="; ymax;
PRINT " "; numberholes; "Holes."
LOCATE 7, 10: PRINT "ID: "; title$: COLOR 0, 7: LOCATE 10, 10
xdest = x(currenthole): ydest = y(currenthole)
IF xdest < 0 OR ydest < 0 THEN BEEP: GOTO autodrill01
a$ = "Moving to Hole ###   (#.##,#.##)"
PRINT USING a$; currenthole; xdest; ydest;
LOCATE 18, 33: COLOR 4, 7: PRINT "Esc To Abort. "; : COLOR 0, 7
which = 128: 'Table Selected - X Direction
xsteps = (xdest - xlocation) * distancecal
ysteps = (ydest - ylocation) * distancecal
IF xsteps < 0 THEN
steps = -1
IF xdest > .1 THEN
```

```
xsteps = ABS(xsteps) + 43
lashflgx = 1
ELSE
xsteps = ABS(xsteps)
lashflgx = 0
END IF
ELSE
steps = 1
lashflgx = 0
END IF
IF xsteps = 0 THEN GOTO doneauto
automove01:
OUT add, which: stepcount = 0
automove02:
aseq = (aseq MOD 4) + steps
IF aseq = 0 THEN aseq = 4
IF aseq = -1 THEN aseq = 3
OUT add, a(aseq) + which
FOR k = 1 TO speed!: NEXT k
stepcount = stepcount + 1
IF (stepcount >= xsteps) AND which = 128 THEN GOTO doneauto
IF (stepcount >= ysteps) AND which = 0 THEN GOTO doneauto
panicstop$ = INKEY$: IF panicstop$ = "" THEN GOTO automove02
IF ASC(panicstop$) = 27 THEN
PCOPY 1, 0: VIEW PRINT: OUT add, 0: OUT add, 128: RETURN
ELSE
GOTO automove02
END IF
doneauto:
IF lashflgx = 1 THEN
steps = 1
xsteps = 43
lashflgx = 0
GOTO automove01
END IF
IF lashflgy = 1 THEN
steps = -1
ysteps = 43
lashflgy = 0
GOTO automove01
END IF
start! = TIMER
```

```
WHILE (TIMER - start!) < .2: WEND
OUT add, 0 + which
IF which = 128 THEN
which = 0: 'Drill Selected - Y Direction
IF ysteps < 0 THEN
steps = 1:
IF ydest > .1 THEN
ysteps = ABS(ysteps) + 43
lashflgy = 1
ELSE
ysteps = ABS(ysteps)
lashflgy = 0
END IF
ELSE steps = -1
END IF
IF ysteps <> 0 THEN GOTO automove01
END IF
xlocation = xdest: ylocation = ydest
IF currenthole < numberholes THEN
LOCATE 12, 10: PRINT "ARRIVED.  Press Any Key for Next Hole..";
a$ = INPUT$(1): IF a$ = CHR$(27) THEN GOTO doneauto01
GOTO autodrill01
END IF
WEND
LOCATE 12, 10: PRINT SPACE$(69); : LOCATE 12, 10
PRINT "DONE.  Press Any Key."; : a$ = INPUT$(1)
doneauto01:
PCOPY 1, 0: VIEW PRINT: OUT add, 0: OUT add, 128: RETURN
'
'INITIAL ALIGNMENT OPTION
'
alignment:
VIEW PRINT 4 TO 25: COLOR 0, 7: CLS : PCOPY 0, 1: COLOR 8, 7
LOCATE 10, 33
PRINT "7": LOCATE 10, 45: PRINT "9": LOCATE 12, 39: PRINT "5"
LOCATE 14, 33: PRINT "1": LOCATE 14, 45: PRINT "3"
FOR i = 10 TO 14: FOR j = 30 TO 48 STEP 6
LOCATE i, j: PRINT CHR$(179);
NEXT j, i
FOR i = 9 TO 16 STEP 2: LOCATE i, 30: PRINT STRING$(18, 196); :
NEXT i
LOCATE 9, 30: PRINT CHR$(218): LOCATE 9, 36: PRINT CHR$(194);
```

```
LOCATE 9, 42: PRINT CHR$(194): LOCATE 9, 48: PRINT CHR$(191);
FOR i = 11 TO 13 STEP 2:
LOCATE i, 30: PRINT CHR$(195): LOCATE i, 36: PRINT CHR$(197)
LOCATE i, 42: PRINT CHR$(197): LOCATE i, 48: PRINT CHR$(180)
NEXT i
LOCATE 15, 30: PRINT CHR$(192): LOCATE 15, 36: PRINT CHR$(193)
LOCATE 15, 42: PRINT CHR$(193): LOCATE 15, 48: PRINT CHR$(217)
COLOR 1, 7
LOCATE 7, 29: PRINT "DRILL TOWARDS STEPPER"
LOCATE 8, 39: PRINT CHR$(24)
LOCATE 12, 3: PRINT "TABLE AWAY FROM STEPPER <—"
LOCATE 12, 49: PRINT "—> TABLE TOWARDS STEPPER"
LOCATE 16, 39: PRINT CHR$(25)
LOCATE 17, 28: PRINT "DRILL AWAY FROM STEPPER"
getdirection:
COLOR 7, 0: LOCATE 20, 7
PRINT " Press Appropriate Keypad Key to Move Table/Drill.";
PRINT "  Esc when Done. ";
COLOR 1, 7: LOCATE 10, 39: PRINT "8": LOCATE 12, 33: PRINT "4"
LOCATE 12, 45: PRINT "6": LOCATE 14, 39: PRINT "2"
a$ = INKEY$: IF a$ = "" THEN GOTO getdirection
adir = ASC(RIGHT$(a$, 1))
COLOR 16, 7
SELECT CASE adir
CASE IS = 27: 'Esc
PCOPY 1, 0: VIEW PRINT: OUT add, 0 + which: RETURN
CASE IS = 56, 72: 'Drill Towards Stepper
LOCATE 10, 39: PRINT "8"
which = 0: steps = -1
CASE IS = 52, 75: 'Table Away From Stepper
LOCATE 12, 33: PRINT "4"
which = 128: steps = 1
CASE IS = 54, 77: 'Table Towards Stepper
LOCATE 12, 45: PRINT "6"
which = 128: steps = -1
CASE IS = 50, 80: 'Drill Away From Stepper
LOCATE 14, 39: PRINT "2"
which = 0: steps = 1
CASE ELSE
BEEP: GOTO getdirection
END SELECT
LOCATE 20, 7: PRINT SPACE$(70)
```

```
COLOR 30, 0: LOCATE 20, 28: PRINT " Press Any Key to Stop."
alignmove01:
OUT add, which: stemp! = speed!
alignmove02:
aseq = (aseq MOD 4) + steps
IF aseq = 0 THEN aseq = 4
IF aseq = -1 THEN aseq = 3
OUT add, a(aseq) + which
FOR k = 1 TO stemp!: NEXT k
IF (INP(add + 1) AND 64) = 0 AND adir = 77 AND which = 128 THEN
GOTO am3
IF (INP(add + 1) AND 128) = 128 AND adir = 80 AND which = 0 THEN
GOTO am3
a$ = INKEY$: IF a$ = "" THEN GOTO alignmove02
'a = ASC(a$): IF a <> 27 THEN GOTO alignmove02
am3:
start! = TIMER
WHILE (TIMER - start!) < .2: WEND
OUT add, 0 + which
GOTO getdirection
'
'START-UP MESSAGE SUBROUTINE (If MESSAGEDISPLAY$<>"NO")
'
startupmessage:
GOSUB optionbox
COLOR 1, 7: LOCATE 8, 15: PRINT STRING$(51, 219)
COLOR 15, 1: LOCATE 8, 33: PRINT "A U T O - X Y": COLOR 0, 7
LOCATE 10, 19: PRINT "(c)1997, JJ Barbarello, Manalapan, NJ
07726"
LOCATE 11, 19: PRINT STRING$(43, 196)
LOCATE 12, 20: PRINT "Access Functions by pressing ALT and the"
LOCATE 13, 20: PRINT "highlighted key. ";
COLOR 1, 7: PRINT "Ex: <ALT> F selects the"
LOCATE 14, 20: PRINT "File function. ";
COLOR 0, 7: PRINT "Access sub-functions by"
LOCATE 15, 20: PRINT "pressing the highlighted key alone. ";
COLOR 1, 7: PRINT "Ex:"
LOCATE 16, 20: PRINT "O selects Open under the File Function."
LOCATE 22, 33: COLOR 8, 7: PRINT "Press Any Key.."; : COLOR 0, 7
a$ = INPUT$(1): RETURN
'
'ERROR TRAPPING SUBROUTINE
```

```
'
errortrap:
IF ERR = 64 THEN
RESUME NEXT
ELSE
FOR i = 1 TO 5
BEEP: start! = TIMER
WHILE (TIMER - start!) < .1: WEND
NEXT i: VIEW PRINT: PCOPY 1, 0: RESUME getkey
END IF
```

It does not use the FINDPORT routine, since the address of the port you want to use is included in the DRLSETUP.DAT file. It does, however, use the same control routines as used in the SPEEDSET program. This is a fairly complex program, containing about 600 lines of program code. As you can see, however, from the previous discussions, only a small portion of that code is dedicated to controlling the electronics that control the stepper motors. The largest share of the code performs formatting, data retrieval, and communication with the user.

One portion of the code that bears some additional discussion performs what we'll call "lash back correction." The drill and table are driven by the single nut lodged in their drivers. Those nuts have some "play" and, therefore, the drill or table can pivot around its driver nut. The guides eliminate most of this potential side-to-side rotation, but the small gap between the guide and the rail still allow some minor movement.

When the drill or table move in one direction (such as x-value increasing or y-value increasing), they stay in the same position relative to the driver nut. It is when they move in the opposite direction that they pivot around their driver nut and move. Without correction, this pivoting would cause a small error in the position of the drill or table.

Lash back correction is a software routine that ensures the drill or table always wind up moving in the same direction to counteract any pivoting. When the software senses that the drill or table are moving in the X-value decreasing (table towards stepper) or Y-value decreasing (drill away from stepper) direction, it adds 43 steps to the travel in that direction. That causes the drill or table to "overshoot" its final position by that 43 steps. Then, it moves the drill or table the same 43 steps in the opposite, *increasing* direction. The result is that the drill and table end up in their intended final position, but they always end up their travel moving in the increasing direction, counteracting any pivoting. Since the opposite movement direction occurs very rapidly, the drill or table appear to "lash back". Thus, the term lash back correction.

The lash back correction is implemented in both the manual and auto drilling subroutines. The following two code segments from the manual drilling subroutine demonstrate how lash back correction is accomplished. In the first code segment, if the movement (xsteps) is negative, the table will be asked to move in the x-value decreasing direction (steps = -1). Then 43 steps are added to the number of steps to be executed (xsteps = ABS(xsteps) + 43). Additionally, the X-direction lash flag (lashflgx) is set to one to identify that lash back correction needs to occur:

```
IF xsteps < 0 THEN
steps = -1
IF xdest > .1 THEN
xsteps = ABS(xsteps) + 43
lashflgx = 1
ELSE
xsteps = ABS(xsteps)
lashflgx = 0
END IF
ELSE
steps = 1
lashflgx = 0
END IF
```

Once the table has stopped its initial movement, the software checks to see if lash back correction needs to be performed (is the lash flag set to 1?). If so, the software sets the movement in the x-value increasing direction (steps = 1), sets the number of steps to move to 43, resets the lash flag to zero, and loops back to perform the movement:

```
IF lashflgx = 1 THEN
steps = 1
xsteps = 43
lashflgx = 0
GOTO manualmove01
```

This same approach is used for the y direction (drill). Lash back correction is not performed in the alignment function.

Wrapping it Up

In this chapter, we've walked through how to use the AUTOXY program. We've also learned how the FINDPORT program identifies and verifies operation of any available parallel port. Finally, we've learned how both SPEEDSET and AUTOXY control the control electronics. In Chapter 11, we'll take a design look at the control electronics (and the power supply). We'll wrap up this project in Chapter 12 with some ideas and program modifications for the stout of heart who would like to add automatic z-axis (drilling) control to *Auto-XY*.

CHAPTER 11
The Control Electronics and Power Supply

Introduction

Although we've put together the control electronics in Chapter 5, and used them to control the *Auto-XY* stepper motors, we haven't taken a close look at the complete circuitry and to see how it works. This is what we'll do in this chapter, as well as review the operation of the power supply.

The Control Electronics Circuitry

Figure 11-1 shows the complete control electronics circuit. It is a standard stepper motor control configuration (transistors Q1-Q12, and associated components), along with a computer-controller relay, K1, which selects either one motor or the other (but not both at the same time). The boxed numbers refer to the parallel port pins. The circuit contains two similar blocks (driven by P1-2 and P1-4), each providing a path from either +5.75 volts or ground to one end of each of the coils. The other two blocks (driven by P1-3 and P1-5) perform a similar function, but their paths travel through relay K1 to the stepper coils.

Let's concentrate on one of these blocks, consisting of R7, R1, Q5, R2, Q1, R5, and Q3. If the signal from P1-5 is high, transistor Q5 conducts and brings the R2/R5 common point near ground. This ground at the base of Q3 forward biases Q3, causing it to conduct, bringing the junction of Q1 and Q3 close to ground. (The voltage across either the MJE2955 or MJE3055 is dependent upon the current being drawn by the load, increasing as the load draws more current.)

Now let's move to the block consisting of R16, Q12, R12, R11, Q8, R14, and Q10. If the signal provided by P1-2 is low, the R11/R14 junction is pulled high through R12. This causes Q8 to conduct, but Q10 to remain off. Note that the A* end of coil A in both stepper motors M1 and M2 are connected to the Q8, Q10 junction. The A end of these coils are connected to relay K1. Thus, K1 can select which of the stepper motors will be energized.

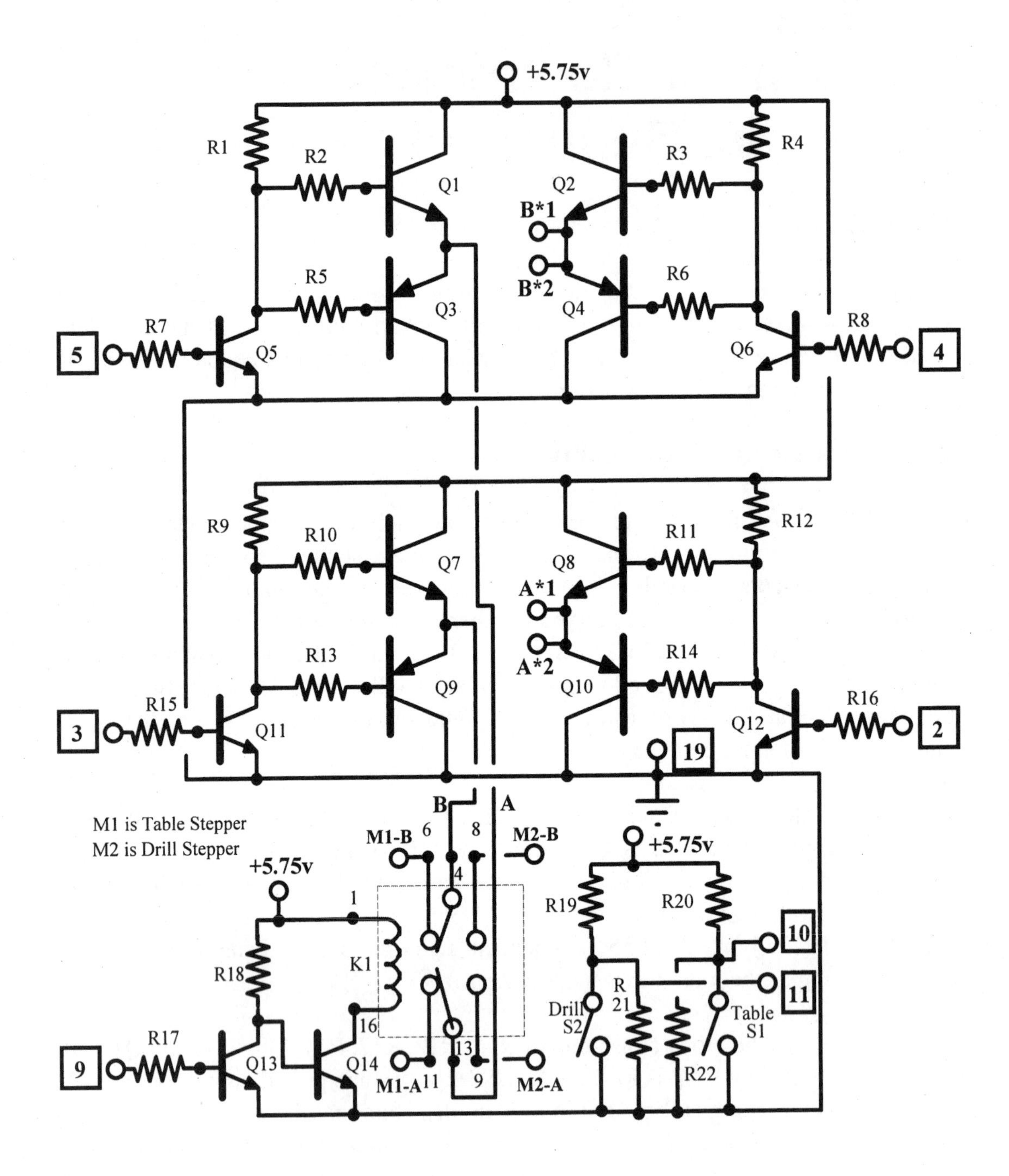

Figure 11-1. The complete control electronics circuit.

If relay K1 is not energized (so the Q1-Q3 junction is connected to stepper motor 1's A coil through K1's pins 11 and 13), a current path will be established through Q8, K1, coil A of stepper motor M1, and Q3. If K1 is energized, the path will now be through Q1, pins 9 and 13 of K1, coil A of stepper motor M2, and Q3.

Transistors Q13 and Q14 form a non-inverting switch that can provide the approximately 100 mA to energize K1 when driven by the low current output of P1-9.

Many a stepper motor control circuit uses a current limiting resistor to guard against excessive current being fed to the motor and causing it to overheat. In this circuit, the relatively large current demand of the motor causes a measurable voltage drop across the power-providing transistors (in the example above, Q3 and Q8). Each will have a base-to-emitter drop approaching one volt. Therefore, the stepper motor coil sees only 3.5 volts or so. With a coil resistance of 6.25 ohms, the current provided is then 3.5/6.25, or about 560 mA. This is well within the motor's ability to dissipate the heat generated by the power being provided to it. Accordingly, no current limiting resistor is required, since sufficient torque is created with this magnitude of current.

The remaining portion of the circuit consists of microswitches S1, S2 and resistor divider strings R19-R21 and R20-R22. In operation, these switches sit untriggered in their normally open state. This provides about 2.9 volts (more than sufficient to be recognized as Logic 1) to P1-10 and P1-11. When either switch is closed, it grounds its associated pin, indicating to the AUTOXY program that the zero position has been reached and the associated stepper motor should be stopped.

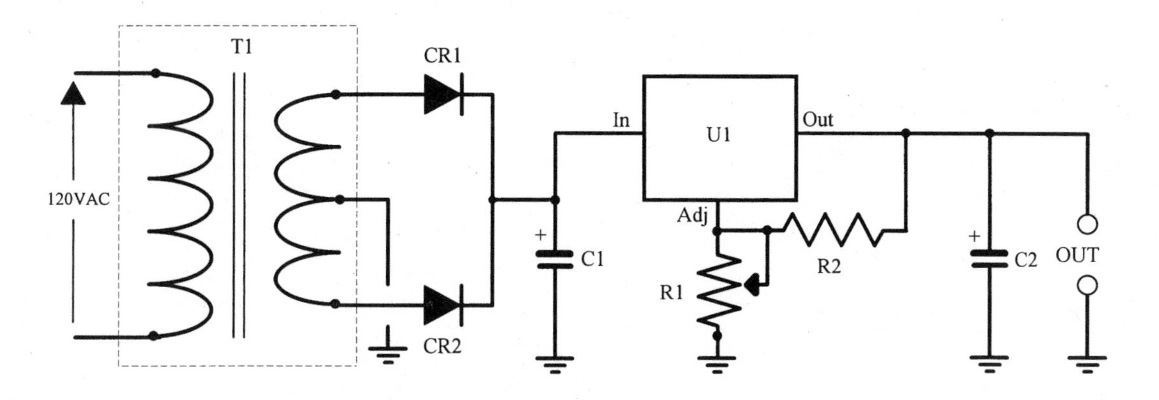

Figure 11-2*. The power supply schematic.*

The Power Supply

The schematic diagram for the power supply is shown in *Figure 11-2*. T1 is a standard transformer with a 120 volt AC primary and a 6.3 volt AC center tapped (or 9 volt AC center tapped) secondary. The secondary's center tap serves as the ground reference point.

The secondary's output is full wave rectified in 1N4003 diodes CR1 and CR2. That full wave rectified AC is smoothed to a low ripple DC by 1000 υF capacitor C1. This DC feeds the LM317T variable voltage regulator U1. Its output is determined by the formula V out = 1.25 (1 + R2/R2). Thus, with a 1,000 ohm potentiometer (R2) and a 100 ohm feedback resistor R1, the output range is 1.25 volts DC to 13.75 volts DC. (Of course, the output cannot be equal to or greater than the input voltage. So this configuration's maximum output voltage will be about a volt below the DC input). The 1 υF capacitor C2 provides output voltage filtering.

CHAPTER 12
Some Thoughts on Adding a Third Axis

Introduction

The *Auto-XY* project was so named because it provides automatic movement in the X and Y directions. So, what about an *Auto-XYZ*, adding automatic movement of the drill up and down? In this final chapter, we'll discuss the factors that make adding a Z-axis a challenge. We'll finish up with some simple software modifications that result in a version called AUTOXYZ (provided on the companion disk), and how you could use it.

The Movement Mechanism

To provide automatic movement in the Z-axis, an electromechanical mechanism needs to be added that either forces the drill down (with a spring bringing it back up), or drives it both up and down (with the ability to hold it in the up position when not drilling). In addition, the mechanism must not force the drill down too abruptly, since any out-of-vertical drill condition could very quickly snap the drill bit. However, the vertical travel needs only to be 1/4" or there about since the drill bit can be positioned as close as a 1/16" above the PC board, and it only has to travel far enough for its cutting tip to clear the bottom of the board.

One possibility is shown in *Figure 12-1.* Here, a supporting frame has been attached to the two drill guides. The plunging mechanism is then attached to the frame so that when it is energized, the drill is pushed downward. A Z-axis limit switch is located in an appropriate position to have it engaged when the drill returns to its resting position, indicating that the hole drilling has been completed.

The plunging mechanism may be electrically operated (like a motor-driven linear shaft), or mechanically operated (like a compressed air-driven cylinder). Regardless of type, the mechanism should not add significant weight to the drill assembly, and should generate sufficient force to move the drill.

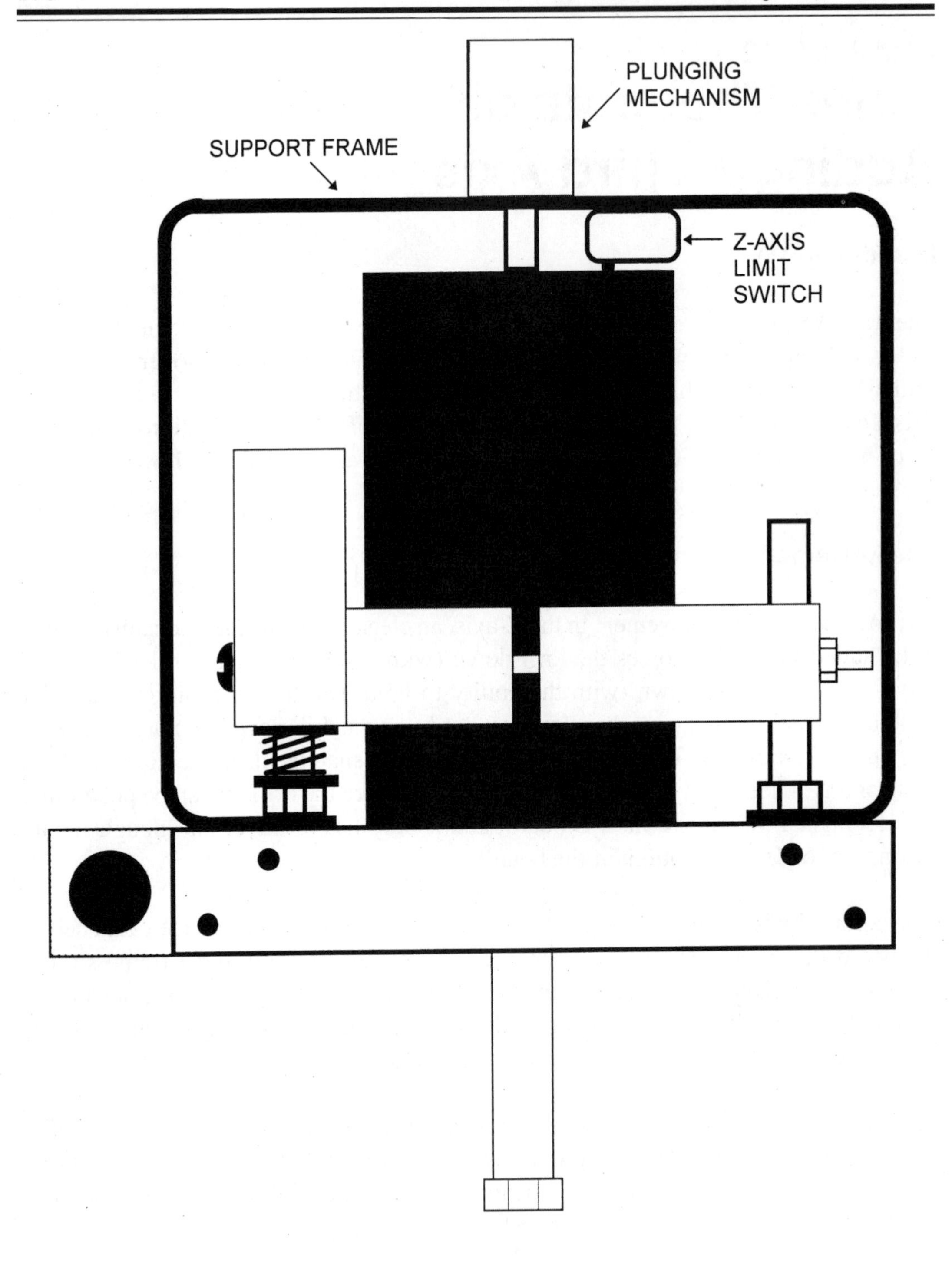

***Figure 12-1**. A supporting frame is attached to the two drill guides.*

Software Control

Modifying the software to allow for automatic Z-axis is relatively easy, requiring the addition of only five lines of code, and the changing of one screen message. These modifications (which are included as bold, underlined text) reside in two subroutines, *GetSetupAndInitSteppers*, and *autodrill*. Rather than list the complete new program (AUTOXYZ), which is essentially the same as AUTOXY with these minor modifications, let's take a look at each affected subroutine individually.

The New Program Code

The first modification is the addition of a single line in the *GetSetupAndInitSteppers* subroutine (see below). When executed, parallel port line 14 (P1-14) is brought low. We will use P1-14 as the output line that signals our plunger mechanism to begin its job of pushing the drill down:

```
'GET INITIALIZATION DATA FROM FILES AND INITIALIZE STEPPERS
'
GetSetupAndInitSteppers:
CLOSE 1: OPEN "DRLSETUP.DAT" FOR INPUT AS #1
LINE INPUT #1, add$: add = VAL(add$)
LINE INPUT #1, speed$: speed! = VAL(speed$)
LINE INPUT #1, distancecal$: distancecal = VAL(distancecal$)
LINE INPUT #1, messagedisplay$
 messagedisplay$ = UCASE$(LEFT$(messagedisplay$, 2))
CLOSE 1
```

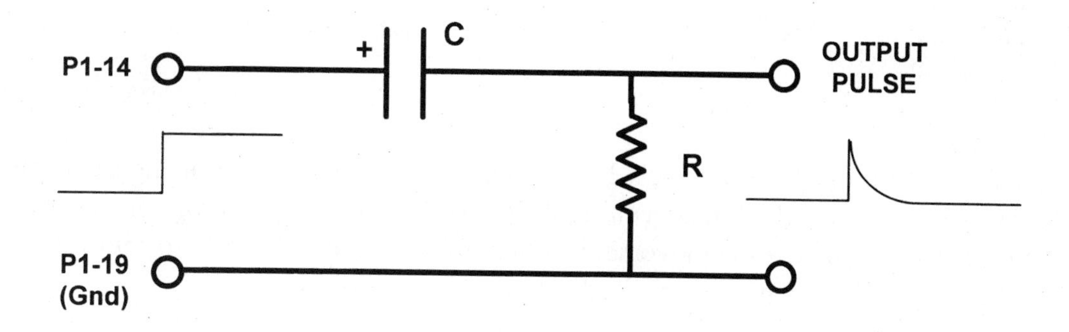

Figure 12-2. *This circuit can be used to convert the P1-14 signal.*

```
OUT add, a(4) + 128: FOR i = 1 TO speed!: NEXT i: OUT add, 128
OUT add, a(4): FOR i = 1 TO speed!: NEXT i: OUT add, 0: aseq = 4
OUT add + 2, 2
RETURN
```

The second modification is made in the *autodrill* subroutine. An abbreviated version of *autodrill* is shown below. The program code between the beginning of this subroutine (*autodrill:*) and the label *doneauto:* remains unchanged from the original AUTOXY. Similarly, the program code between the *doneauto:* label and the line *xlocation* = *xdest: ylocation* = *ydest* also remains unchanged. After this line, four new lines have been inserted:

```
'AUTO DRILLING SUB-OPTION
'
autodrill:
. . .
doneauto:
. . .
xlocation = xdest: ylocation = ydest
OUT add + 2, 0
IF currenthole < numberholes THEN
LOCATE 12, 10: PRINT "ARRIVED.  Trigger Pin 12 for Next Hole..";
start! = TIMER
WHILE (TIMER - start!) < 1: WEND
WHILE (INP(add + 1) AND 32) = 32: WEND: OUT add + 2, 2
GOTO autodrill01
END IF
WEND
LOCATE 12, 10: PRINT SPACE$(69); : LOCATE 12, 10
PRINT "DONE.  Press Any Key."; : a$ = INPUT$(1)
doneauto01:
PCOPY 1, 0: VIEW PRINT: OUT add, 0: OUT add, 128: RETURN
```

The first new program line brings P1-14 high. The next program change is in the "Arrived" message, which has been modified to indicate that the program will now wait for the automatic drill movement to be completed, and the drill to return to its resting position before proceeding to the next hole location.

The next two new program lines cause a delay of one second to allow the drill to move from its resting position. When the drill moves, it disengages from the z-axis limit switch. If we did not wait, the plunger mechanism would be energized at the same time the program looked to see if the drill had "returned" from being plunged.

The final new program line continuously checks the status of P1-12, looking for a logic zero. When that is found, the line returns P1-14 to a low state, ready for the next hole.

Using the Output Trigger Signal

P1-14 outputs a constant logic high when a hole location has been reached. That signal can serve as the input to subsequent circuitry that drives a motor, energizes a relay, or otherwise causes the plunger mechanism to operate. If a pulse (rather than a continuous signal) is desired, the simple circuit of *Figure 12-2* can be used to convert the P1-14 signal. Start with values of 1 υF for C, and 10 kohms for R. Adjust these values as necessary to obtain the appropriate level and duration pulse for your subsequent circuitry.

Wrapping it Up

With this discussion, we've come to the end of this project. It's time for you to get out your CAD program (or a paper and pencil) and begin laying out a PC Board for your latest creation.

If you have any questions, or would like to discuss ideas or applications, jot them down and include them with a self-addressed stamped envelope to:

JJ Barbarello
817 Tennent Road
Manalapan, NJ 07726

Happy Experimenting!

INDEX

SYMBOLS

A

B

D

E

T

Alternative Energy

Mark E. Hazen

This book is designed to introduce readers to the many different forms of energy mankind has learned to put to use. Generally, energy sources are harnessed for the purpose of producing electricity. This process relies on transducers to transform energy from one form into another. *Alternative Energy* will not only address transducers and the five most common sources of energy that can be converted to electricity, it will also explore solar energy, the harnessing of the wind for energy, geothermal energy, and nuclear energy.

This book is designed to be an introduction to energy and alternate sources of electricity. Each of the nine chapters are followed by questions to test comprehension, making it ideal for students and teachers alike. In addition, listings of World Wide Web sites are included so that readers can learn more about alternative energy and the organizations devoted to it.

Professional Reference
320 pages • Paperback • 7-3/8 x 9-1/4"
ISBN: 0-7906-1079-5 • Sams: 61079
$18.95 ($25.95 Canada) • October 1996

The Complete RF Technician's Handbook

Cotter W. Sayre

The *Complete RF Technician's Handbook* will furnish the working technician or student with a solid grounding in the latest methods and circuits employed in today's RF communications gear. It will also give readers the ability to test and troubleshoot transmitters, transceivers, and receivers with absolute confidence. Some of the topics covered include reactance, phase angle, logarithms, diodes, passive filters, amplifiers, and distortion. Various multiplexing methods and data, satellite, spread spectrum, cellular, and microwave communication technologies are discussed.

Cotter W. Sayre is an electronics design engineer with Goldstar Development, Inc., in Lake Elsinore, California. He is a graduate of Los Angeles Pierce College and is certified by the National Association of Radio and Telecommunications Engineers, as well as the International Society of Electronics Technicians.

Professional Reference
281 pages • Paperback • 8-1/2 x 11"
ISBN: 0-7906-1085-X • Sams: 61085
$24.95 ($33.95 Canada) • July 1996

CALL 1-800-428-7267 TODAY FOR THE NAME OF YOUR NEAREST PROMPT PUBLICATIONS DISTRIBUTOR

PROMPT
PUBLICATIONS

ES&T Presents TV Troubleshooting & Repair

Electronic Servicing & Technology Magazine

TV set servicing has never been easy. The service manager, service technician, and electronics hobbyist need timely, insightful information in order to locate the correct service literature, make a quick diagnosis, obtain the correct replacement components, complete the repair, and get the TV back to the owner.

ES&T Presents TV Troubleshooting & Repair presents information that will make it possible for technicians and electronics hobbyists to service TVs faster, more efficiently, and more economically, thus making it more likely that customers will choose not to discard their faulty products, but to have them restored to service by a trained, competent professional.

Originally published in *Electronic Servicing & Technology*, the chapters in this book are articles written by professional technicians, most of whom service TV sets every day.

Video Technology
226 pages • Paperback • 6 x 9"
ISBN: 0-7906-1086-8 • Sams: 61086
$18.95 ($25.95 Canada) • August 1996

ES&T Presents Computer Troubleshooting & Repair

Electronic Servicing & Technology

ES&T is the nation's most popular magazine for professionals who service consumer electronics equipment. PROMPT® Publications, a rising star in the technical publishing business, is combining its publishing expertise with the experience and knowledge of *ES&T's* best writers to produce a new line of troubleshooting and repair books for the electronics market. Compiled from articles and prefaced by the editor in chief, Nils Conrad Persson, these books provide valuable, hands-on information for anyone interested in electronics and product repair.

Computer Troubleshooting & Repair is the second book in the series and features information on repairing Macintosh computers, a CD-ROM primer, and a color monitor. Also included are hard drive troubleshooting and repair tips, computer diagnostic software, networking basics, preventative maintenance for computers, upgrading, and much more.

Computer Technology
288 pages • Paperback • 6 x 9"
ISBN: 0-7906-1087-6 • Sams: 61087
$18.95 ($26.50 Canada) • February 1997

CALL 1-800-428-7267 TODAY FOR THE NAME OF YOUR NEAREST PROMPT PUBLICATIONS DISTRIBUTOR

Speakers for Your Home & Automobile

Gordon McComb, Alvis J. Evans, & Eric J. Evans

The cleanest CD sound, the quietest turntable, or the clearest FM signal are useless without a fine speaker system. This book not only tells readers how to build quality speaker systems, it also shows them what components to choose and why. The comprehensive coverage includes speakers, finishing touches, construction techniques, wiring speakers, and automotive sound systems.

Gordon McComb has written over 35 books and 1,000 magazine articles which have appeared in such publications as *Popular Science*, *Video*, *Omni*, *Popular Electronics*, and *PC World*. His writings has spanned a wide range of subjects, from computers, to video, to robots. Alvis and Eric Evans are the co-authors of many books and articles on the subject of electricity and electronics. Alvis is also an Associate Professor of Electronics at Tarrant County Junior College in Ft. Worth, Texas.

Audio Technology
164 pages • Paperback • 6 x 9"
ISBN: 0-7906-1025-6 • Sams: 61025
$14.95 ($20.95 Canada) • November 1992

Sound Systems for Your Automobile

Alvis J. Evans & Eric J. Evans

This book provides the average vehicle owner with the information and skills needed to install, upgrade, and design automotive sound systems. From terms and definitions straight up to performance objectives and cutting layouts, *Sound Systems* will show the reader how to build automotive sound systems that provide occupants with live performance reproductions that rival home audio systems.

Whether starting from scratch or upgrading, this book uses easy-to-follow steps to help readers plan their system, choose components and speakers, and install and interconnect them to achieve the best sound quality possible. Installations on specific types of vehicles are discussed, including separate chapters on coupes and sedans, hatchbacks, pickup trucks, sport utility vehicles, and vans. Alvis J. Evans is the author of many books on the subjects of electricity and electronics for both beginning hobbyists and advanced technicians.

Audio Technology
124 pages • Paperback • 6 x 9"
ISBN: 0-7906-1046-9 • Sams: 61046
$16.95 ($22.99 Canada) • January 1994

CALL 1-800-428-7267 TODAY FOR THE NAME OF YOUR NEAREST PROMPT PUBLICATIONS DISTRIBUTOR

The Video Book

Gordon McComb

Televisions and video cassette recorders have become part of everyday life, but few people know how to get the most out of these home entertainment devices. *The Video Book* offers easy-to-read text and clearly illustrated examples to guide readers through the use, installation, connection, and care of video system components. Simple enough for the new buyer, yet detailed enough to assure proper connection of the units after purchase, this book is a necessary addition to the library of every modern video consumer. Topics included in the coverage are the operating basics of TVs, VCRs, satellite systems, and video cameras; maintenance and troubleshooting; and connectors, cables, and system interconnections.

Gordon McComb has written over 35 books and 1,000 magazine articles, which have appeared in such publications as *Popular Science*, *Video*, *PC World*, and *Omni*, as well as many other top consumer and trade publications.

Video Technology

192 pages • Paperback • 6 x 9"
ISBN: 0-7906-1030-2 • Sams: 61030
$16.95 ($22.99 Canada) • October 1992

TV Video Systems

L.W. Pena & Brent A. Pena

Knowing which video programming source to choose, and knowing what to do with it once you have it, can seem overwhelming. Covering standard hard-wired cable, large-dish satellite systems, and DSS, *TV Video Systems* explains the different systems, how they are installed, their advantages and disadvantages, and how to troubleshoot problems. This book presents easy-to-understand information and illustrations covering installation instructions, home options, apartment options, detecting and repairing problems, and more. The in-depth chapters guide you through your TV video project to a successful conclusion.

L.W. Pena is an independent certified cable TV technician with 14 years of experience who has installed thousands of TV video systems in homes and businesses. Brent Pena has eight years of experience in computer science and telecommunications, with additional experience as a cable installer.

Video Technology

124 pages • Paperback • 6 x 9"
ISBN: 0-7906-1082-5 • Sams: 61082
$14.95 ($20.95 Canada) • June 1996

CALL 1-800-428-7267 TODAY FOR THE NAME OF YOUR NEAREST PROMPT PUBLICATIONS DISTRIBUTOR

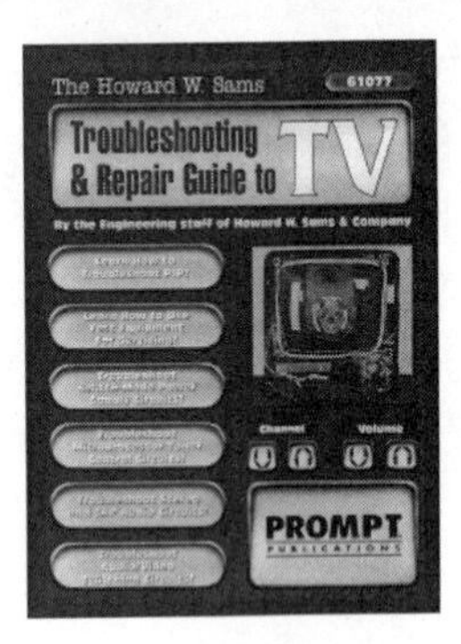

The Howard W. Sams Troubleshooting & Repair Guide to TV

Howard W. Sams & Company

The Howard W. Sams Troubleshooting & Repair Guide to TV is the most complete and up-to-date television repair book available. Included in its more than 300 pages is complete repair information for all makes of TVs, timesaving features that even the pros don't know, comprehensive basic electronics information, and extensive coverage of common TV symptoms.

This repair guide is completely illustrated with useful photos, schematics, graphs, and flowcharts. It covers audio, video, technician safety, test equipment, power supplies, picture-in-picture, and much more. *The Howard W. Sams Troubleshooting & Repair Guide to TV* was written, illustrated, and assembled by the engineers and technicians of Howard W. Sams & Company. This book is the first truly comprehensive television repair guide published in the 90s, and it contains vast amounts of information never printed in book form before.

Video Technology
384 pages Paperback • 8-1/2 x 11"
ISBN: 0-7906-1077-9 • Sams: 61077
$29.95 ($39.95 Canada) • June 1996

The In-Home VCR Mechanical Repair & Cleaning Guide

Curt Reeder

Like any machine that is used in the home or office, a VCR requires minimal service to keep it functioning well and for a long time. However, a technical or electrical engineering degree is not required to begin regular maintenance on a VCR. *The In-Home VCR Mechanical Repair & Cleaning Guide* shows readers the tricks and secrets of VCR maintenance using just a few small hand tools, such as tweezers and a power screwdriver.

This book is also geared toward entrepreneurs who may consider starting a new VCR service business of their own. The vast information contained in this guide gives a firm foundation on which to create a personal niche in this unique service business. This book is compiled from the most frequent VCR malfunctions Curt Reeder has encountered in the six years he has operated his in-home VCR repair and cleaning service.

Video Technology
222 pages • Paperback • 8-3/8 x 10-7/8"
ISBN: 0-7906-1076-0 • Sams: 61076
$19.95 ($26.99 Canada) • April 1996

CALL 1-800-428-7267 TODAY FOR THE NAME OF YOUR NEAREST PROMPT PUBLICATIONS DISTRIBUTOR

Internet Guide to the Electronics Industry

John Adams

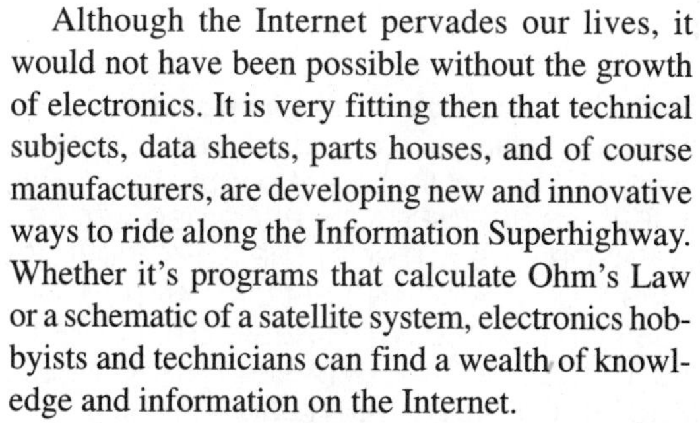

Although the Internet pervades our lives, it would not have been possible without the growth of electronics. It is very fitting then that technical subjects, data sheets, parts houses, and of course manufacturers, are developing new and innovative ways to ride along the Information Superhighway. Whether it's programs that calculate Ohm's Law or a schematic of a satellite system, electronics hobbyists and technicians can find a wealth of knowledge and information on the Internet.

In fact, soon electronics hobbyists and professionals will be able to access on-line catalogs from manufacturers and distributors all over the world, and then order parts, schematics, and other merchandise without leaving home. The *Internet Guide to the Electronics Industry* serves mainly as a directory to the resources available to electronics professionals and hobbyists.

Internet
192 pages • Paperback • 5-1/2 x 8-1/2"
ISBN: 0-7906-1092-2 • Sams: 61092
$16.95 ($22.99 Canada) • December 1996

Harmonics

Mark Waller

Harmonics is the essential guide to understanding all of the issues and areas of concern surrounding harmonics and the recognized methods for dealing with them.

Covering nonlinear loads, multiple PCs, K-factor transformers, and more, Mark Waller prepares the reader to manage problems often encountered in electrical distribution systems that can be solved easily through an understanding of harmonics, current, and voltage.

This book is a useful tool for system and building engineers, electricians, maintenance personnel, and all others concerned about protecting and maintaining the quality of electrical power systems.

Electrical Technology
132 pages • Paperback • 7-3/8 x 9-1/4"
ISBN: 0-7906-1048-5 • Sams: 61048
$24.95 ($33.95 Canada) • May 1994

CALL 1-800-428-7267 TODAY FOR THE NAME OF YOUR NEAREST PROMPT PUBLICATIONS DISTRIBUTOR

AGREEMENT

READ THIS AGREEMENT BEFORE OPENING THE SOFTWARE PACKAGE

BY OPENING THE SEALED PACKAGE YOU ACCEPT AND AGREE TO THE FOLLOWING TERMS AND CONDITIONS PRINTED BELOW. IF YOU DO NOT AGREE, DO NOT OPEN THE PACKAGE AND RETURN THE SEALED PACKAGE AND ALL MATERIALS YOU RECEIVED TO HOWARD W. SAMS & COMPANY, 2647 WATERFRONT PARKWAY EAST DRIVE SUITE 300, INDIANAPOLIS, IN 46214-2041 (HEREINAFTER "LICENSOR") WITHIN 30 DAYS OF RECEIPT ALONG WITH PROOF OF PAYMENT.

Licensor retains the ownership of this copy and any subsequent copies of the Software. This copy is licensed to you for use under the following conditions:

Permitted Uses. You may: use the Software on any supported computer configuration, provided the Software is used on only one such computer and by one user at a time; permanently transfer the Software and its documentation to another user, provided you retain no copies and the recipient agrees to the terms of this Agreement.

Prohibited Uses. You may not: transfer, distribute, rent, sub-license, or lease the Software or documentation, except as provided herein; alter, modify, or adapt the Software or documentation, or portions thereof including, but not limited to, translation, decompiling, disassembling, or creating derivative works; make copies of the documentation, the Software, or portions thereof; export the Software.

LIMITED WARRANTY, DISCLAIMER OF WARRANTY

Licensor warrants that the optical media on which the Software is distributed is free from defects in materials and workmanship. Licensor will replace defective media at no charge, provided you return the defective media with dated proof of payment to Licensor within ninety (90) days of the date of receipt. This is your sole and exclusive remedy for any breach of warranty. EXCEPT AS SPECIFICALLY PROVIDED ABOVE, THE SOFTWARE IS PROVIDED ON AN "AS IS" BASIS, LICENSOR, THE AUTHOR, THE SOFTWARE DEVELOPERS, PROMPT PUBLICATIONS, HOWARD W. SAMS & COMPANY, AND BELL ATLANTIC MAKE NO WARRANTY OR REPRESENTATION, EITHER EXPRESS OR IMPLIED, WITH RESPECT TO THE SOFTWARE, INCLUDING ITS QUALITY, ACCURACY, PERFORMANCE, MERCHANTABILITY, OR FITNESS FOR A PARTICULAR PURPOSE. IN NO EVENT WILL LICENSOR, THE AUTHOR, THE SOFTWARE DEVELOPERS, PROMPT PUBLICATIONS, HOWARD W. SAMS & COMPANY, AND BELL ATLANTIC BE LIABLE FOR DIRECT, INDIRECT, SPECIAL, INCIDENTAL, OR CONSEQUENTIAL DAMAGES (INCLUDING, BUT IS NOT LIMITED TO, INTERRUPTION OF SERVICE, LOSS OF DATA, LOSS OF CLASSROOM TIME, LOSS OF CONSULTING TIME) OR LOST PROFITS ARISING OUT OF THE USE OR INABILITY TO USE THE SOFTWARE OR DOCUMENTATION, EVEN IF ADVISED OF THE POSSIBILITY OF SUCH DAMAGES. IN NO CASE SHALL LIABILITY EXCEED THE AMOUNT OF THE FEE PAID. THE WARRANTY AND REMEDIES SET FORTH ABOVE ARE EXCLUSIVE AND IN LIEU OF ALL OTHERS, ORAL OR WRITTEN, EXPRESS OR IMPLIED. Some states do not allow the exclusion or limitation of implied warranties or limitation of liability for incidental or consequential damages, so that the above limitation or exclusion may not apply to you.

GENERAL:

Licensor retains all rights, not expressly granted herein. This Software is copyrighted, nothing in this Agreement constitutes a waiver of Licensor's rights under United States copyright law. This license is non-exclusive. This License and your right to use the Software automatically terminate without notice from Licensor if you fail to Comply with any provision of this Agreement. This Agreement is governed by the laws of the State of Indiana.